A Guide to UK and Ireland Electric Vehicle Charge Points

A N Hurst

Joe Public

Copyright © 2015 by Andrew Hurst

All rights reserved. No part of this publication may be reproduced, distributed, or transmitted in any form or by any means, including photocopying, recording, or other electronic or mechanical methods, without the prior written permission of the publisher. The only exception is by a reviewer, who may quote short excerpts in a review.

Cover design by Studio Aitch
All pictures by Studio Aitch unless marked otherwise
Published by Joe Public Publishing

Publisher's Cataloguing–in–Publication data
Hurst, A N.
A title of book: *A Guide to UK and Ireland Electric Vehicle Charge Points* / A N Hurst

The names and/or reference to any manufacturers, suppliers, products, services, brands, models etc appearing in this book are the trademarks or registered trademarks of their respective owners. They are used for illustrative purposes only and do not imply any endorsement. Pictures/drawings shown are not contractual and not to scale.

E&OE

First Edition Paperback December, 2015
Second Edition Paperback February, 2016
Information correct to 6th February, 2016

ISBN–10: 151955575X
ISBN–13: 978-1519555755

Front Cover/Previous Page: *A Vauxhall Ampera on charge at a GMEV charge point at Manchester Central, Manchester.*

Rear Cover Top: *A Source West branded, CYC operated Type 2 charge point at Odd Down Park and Ride, Bath.* ***Bottom:*** *A Tesla Model S at a Supercharger, Warrington.*

Please Note!

This book refers to the electric vehicle charging infrastructure in the UK and Republic of Ireland only. For clarity, electric vehicle charging infrastructure means charging points owned and/or operated by third party commercial organisations, as opposed to any domestic or privately installed charging points. By UK and Republic of Ireland, I am referring to:

- England
- Wales
- Scotland
- Northern Ireland
- Republic of Ireland

Plus any outlying islands including the Isle of Wight, the Isle of Mann, Isle of Anglesey and the various Scottish islands that may have charging infrastructure installed under the various schemes detailed within this book.

The book is useful if you:

- Own or operate an Electric Vehicle (EV – car, van, truck, motorbike or scooter), including an Extended Range Electric Vehicle (range extender or ER–EV / EREV) or a plug in hybrid (the vehicle can be recharged from an external fixed power supply and has a petrol engine as well). For clarity I will refer to anything that can be recharged from an external power supply as an Electric Vehicle or EV.
- Intend to buy any of the above vehicle types in the UK
- Intend to visit the UK with any of the above vehicle types
- Have a general interest in the current state of the UK electric vehicle charging infrastructure

Date Information Is Correct To

The date the information is correct to and the publishing date of the book are given at the bottom of page 2. This book will be updated as and when relevant. The latest version will be available on all good book selling websites.

Information Updates

If you have any information that can be used to update this book, then please let me know, so I can add it in to the next issue. You can contact me via Twitter @ANHurst or via the Joe Public Publishing Facebook Page.

Contents

Please Note!   3

1.0   Introduction   8
     1.1   Market Fragmentation   8
     1.2   Plugged–in–Places   9
     1.3   Government Intervention   9
     1.4   Market Fragmentation   10
     1.5   Charge Point Access Cards   11
     1.6   Things Can Only Get Better…   11

2.0   EV Benefits   13

3.0   Electric Vehicles   14
     3.1   Electric Vehicle   14
     3.2   Hybrid/Extended Range Electric Vehicles   15

4.0   Recharging Costs   16
     4.1   Electricity Costs   16
     4.2   Paying for Charge Scheme Electricity   17

5.0   Car Charging Sockets   18
     5.1   AC Car Charger Socket   18
     5.2   DC Car Charger Socket   19

6.0   Charging Connector Types   20
     6.1   AC Fixed Power Supply/Charger Connectors   20
     6.2   DC Fixed Power Supply/Charger Connectors   20

7.0   Charge Point Types   21
7.1   Slow Charge Points   21
7.2   Fast Charge Points   22
         7.2.1   Fast Type 1 Vehicle Connectors   23
         7.2.2   Fast Type 2 Vehicle Connectors   24
     7.3   Rapid Charge Points   24
         7.3.1   Rapid AC Charge Points   25
             7.3.1.1   Electric Vehicle   25
             7.3.1.2   Hybrid/Extended Range Electric Vehicle
                                         25

		7.3.2	Rapid DC Chargers	27
		7.3.3	Why are there two versions of Rapid Charger?	27
		7.3.4	Tesla Supercharger	29
		7.3.5	Tri–charger or Triplex Rapid Charger	31
		7.3.6	UltraCharger	31
	7.4	Charge Point Back Office Standards		31
	7.5	Charging Modes		32
8.0	Joining an Electric Vehicle Charging Scheme			33
	8.1	Charge Point Location Maps		33
9.0	Charge Point Use			35
	9.1	Using Your Own Charge Cable		35
	9.2	Using Your Own Charge Cable at a Public Charge Point		
				36
	9.3	Using a Tethered Charge Cable on a Rapid Charge Point		
				37
	9.4	Rapid Chargers and Heart Conditions		39
	9.5	ICED		39
10.0	Charge Point Scheme Providers			42
	10.1	ChargePoint Genie		44
		10.1.1	Details	44
	10.2	ChargerNet		46
		10.2.1	Details	46
	10.3	Charge Your Car		48
		10.3.1	Details	48
		10.3.2	CYC Access Card Costs	48
		10.3.3	CYC App	48
		10.3.4	Telephone Use	48
		10.3.5	Top Tip!	48
	10.4	ecarNI		49
		10.4.1	Details	49
		10.4.2	Roaming	49
	10.5	Ecotricity		50
		10.5.1	Details	50
	10.6	Electricity Corby		52
		10.6.1	Details	52
	10.7	Energise Network		53
		10.7.1	Details	53

10.8	ESB ecars		54
	10.8.1 Details		54
	10.8.2 Roaming		54
10.9	GMEV – Greater Manchester Electric Vehicle		56
	10.9.1 Details		56
10.10	Greener Scotland / ChargePlace Scotland		58
	10.10.1 Details		58
10.11	Hackney		59
	10.11.1 Details		59
10.12	inmotion!		60
	10.12.1 Details		60
10.13	Just Park		61
	10.13.1 Details		61
10.14	Milton Keynes Crosslink		63
	10.14.1 Details		63
10.15	Plugged–in–Midlands		64
	10.15.1 Details		64
10.16	POD Point		65
	10.16.1 Details		65
	10.16.2 POD Point Open Charge		65
10.17	Polar		67
	10.17.1 Details		67
10.18	Recharge		68
	10.18.1 Details		68
10.19	Source East		70
	10.19.1 Details		70
	10.19.2 Roaming		71
10.20	Source London		72
	10.20.1 Details		72
10.21	Source West		73
	10.21.1 Details		73
10.22	SSE		74
	10.22.1 Details		74
10.23	Tesla Supercharger Network		76
	10.23.1 Details		76
10.24	Think Travel/Think Electric		77
	10.24.1 Details		77
10.25	ZeroNet Network by Zero Carbon World		78
	10.25.1 Details		78

11.0	Businesses with Charge Points			80
12.0	The Future?			82
	12.1	Induction Charging		82
	12.2	Battery Swapping		82
	12.3	Solar		83
	12.4	Other Low Carbon Emission Technologies		83
		12.4.1	Hydrogen Fuel Cell	83
		12.4.2	LPG/Petrol	85
		12.4.3	Hydrogen/Diesel	86
		12.4.4	Bio-methane	86
	12.5	Hindsight: A Wonderful Thing		87
	12.6	So what is the ultimate EV Charge Point?		87
		12.6.1	Standardisation of vehicle charge point socket	87
		12.6.2	Longer Battery Range, Faster Battery Recharge Times and Lower Battery Cost	88
		12.6.3	Variable Recharge Times for EVs	88
		12.6.4	The right sort of charger at the right place	88
		12.6.5	24/7 Charger Access	89
		12.6.6	Standardised Electricity Charges	89
		12.6.7	A variety of Charge Point Providers	89
		12.6.8	Standardised Payment Format	90
		12.6.9	Charge Point Connections	90
		12.6.10	Petrol Stations to become Energy Stations	90
		12.6.11	More Car Park Charge Points	91
13.0	About the Author			92
14.0	Acknowledgements			93
15.0	One Last Thing!			93

Appendices

Appendix A – Useful Websites	94
Appendix B – EV/EREV Vehicle Charge Socket List	96
Appendix C – Glossary of Terms	98

1.0 Introduction

I bought my Extended Range Electric Vehicle (EREV) in April 2014. Due to where I live in the UK, the type of trips I tend to do and the then current state of the electric vehicle charging infrastructure in the UK, I opted to purchase a Vauxhall Ampera. This is basically the UK version of the US manufactured Chevrolet Volt. As an EREV and not a pure electric vehicle, the Ampera has a 1.4 litre petrol Internal Combustion Engine (ICE) which allowed it to continue on its way when the battery was depleted ("empty") and would not leave me stranded at the side of the road.

The Ampera was primarily an electric car with an ICE, unlike the original Prius which was an ICE with an electric motor and battery, which was able to recover breaking energy. The Ampera "wanted" to be an electric vehicle, seemed to drive better as an electric vehicle and was of course much more economical per mile as an electric vehicle. I had calculated that on 100% battery and charging at home, driving cost me around 4p a mile, whereas on 100% petrol it cost me about 8p a mile. This was based on November 2015 energy prices, with unleaded petrol at £1.037/litre.

For this reason and the fact that I had bought an EREV, I was keen to be recharging the car **at every opportunity I had**, to maximise electric driving. This of course was to prove very challenging in practise and is one of the reasons I have decided to write this book, to share my learned knowledge on what charging infrastructure is available.

1.1 Market Fragmentation

One of the issues with electric vehicles is the old "catch 22" chicken and egg scenario. Electric vehicles are only good if there is electric vehicle charge point infrastructure to recharge them. Charge points are only good if there are electric vehicles to use them, especially if they are installed by commercial enterprises, wanting to make a return on their financial investment. Left to market forces and the usual supply and demand economics, the kick off of electric vehicles and the associated infrastructure looked uncertain.

Fortunately the UK Government, through the requirements of cutting carbon emissions and the usual environmental pressures placed upon them at an international level, stepped in with some money to help kick start the infrastructure installation process and the purchase of EVs through fund matching via a scheme called "Plugged–in–Places". Money allowed home owners, local councils and corporate businesses the opportunity to install the necessary charge point equipment to move things along and build the "electric highway". Private individuals could also get up to £5000 off the price of a new car purchase via the Plug in Car Grant. This kick start was undertaken in 2009 via the Office for Low Emission Vehicles (OLEV) which is part of the UK Government's Department for Transport, Department for Business, Innovation and Skills and Department of Energy and Climate Change.

In the Republic of Ireland, similar government back funding via the EU was used to kick start the EV charge point infrastructure roll out. Being a geographically smaller country, the process was somewhat simpler, with only one main player being involved.

1.2 Plugged–in–Places

OLEV's Plugged–in–Places scheme offers fund matching to business consortiums and public sector organisations to install electric vehicle charge points in a particular area. The money has to be bid for in a competition against other schemes, and has to detail the economic benefit it will bring to the area in the form of a cost/benefit analysis, as well as where the charge points will be installed and what type.

Of course, Plugged–in–Places is not a one way street (pun intended) of funding without a return. Data from successful schemes must be provided back to OLEV on how drivers are using the scheme to recharge their electric vehicles, including when, how long, type of vehicle etc. This is to provide data to help OLEV create a UK network of recharging points that meets the needs of EV drivers, including greater interoperability, better accessibility and the means to allow longer journeys. Similar data is also transmitted by your home charge point, which typically was installed without cost to you via government funding, every time you plug in to charge your EV. Big Brother is indeed watching you!

Initially there were 8 Plugged–in–Places schemes across the UK which all received a portion of the £30 million fund. This has since been expanded to further schemes with additional funds.

1.3 Government Intervention

Whilst Government intervention into the market place is a very socialist thing to do for a Conservative led Coalition Government (at the time), the actual method to distribute the money that was available was undertaken in a free market manner. So instead of having some joined up UK wide thinking for an electric highway that was open to anyone with an EV, what has been built is the exact opposite. This is of course at odds to what OLEV are trying to promote. There is not one overarching system provider, but a plethora of individual, stand alone organisations, each with their own ideas and ways of doing things, including charge point membership requirements and electricity prices.

Whilst the actual charger sockets on the vehicles and the charging infrastructure conform to the relevant national and international electrical requirements and standards, the actual roadside equipment typically differs between the charger organisations who have installed it. In addition, the types of charger differ between locations, meaning not all cars can be recharged at all locations. Also, unless you have the relevant access card for the relevant charging scheme, you cannot just turn up and use a charger easily, if at all. This is an absolute nonsense, meaning you potentially need an access card for each and every scheme you may want to use, or have the hassle of making telephone calls and quoting bank card numbers to call centres instead. Not a very attractive proposition, when you consider the ease with which you can drive into **any** petrol station and put fuel in your vehicle.

1.4   Market Fragmentation

This fragmented approach means that as well as the requirement to have an access card for each and every scheme you may use (unless there is a pre-agreed access sharing agreement between schemes, which there aren't that many yet), you typically have to pay for each scheme's card up front! The schemes all charge varying amounts per card, from nothing to anything up to £22 **a year**. On top of that, the amount charged for the electricity varies greatly between schemes, from free electricity to a charge per kWh that it more expensive than running a petrol car for the equivalent miles!!!

All of this is very unhelpful for the EV driver and, as mentioned above does not provide the same ease of refuelling as an ICE driver enjoys by driving into any petrol station, anywhere in the country. You don't have certain petrol stations that only cater for certain car types, even though there are a variety of different brands of petrol station in existence. Nor do you have to pre–register for a certain brand of petrol station if you are going to be travelling out of your usual area or pay an up front, yearly fee, to be able to use a certain petrol station. But for EV drivers this is basically what the government has allowed to happen in the UK, which unfortunately makes it all very complicated and unattractive, especially if you have to travel nationally every so often, as most of us do!

However, some good news is that in recent months it would seem that some of the schemes have started to allow access to other scheme provider's card holders. For instance Ecotricity is now allowing access by card holders of Source London, Plugged–in Midlands, Source East and Charge Your Car to certain charge points on the motorway that fall within these charge providers geographical areas. Complicated? You bet!

1.5   Charge Point Access Cards

The access card type used by the most charging schemes is what is technically known as RFID. This stands for Radio Frequency Identification and is basically the same as your contactless bank card that you may have in your wallet or purse now. The card has a built in passive loop antenna which is detected when placed on or near an active RFID reader. The reader, which sends out a low powered radio frequency, "sees" the card's identification number and compares this to a database it holds within the charging point. If your name is down on the list as having a valid card/account, then you can use the charging point.

As with any bank or access card, care must be taken with it. Handle it with care and be careful not to crack it through bending it. As the card does not have a magnetic stripe, you don't need to worry about it coming into contact with magnetic sources, like speaker magnets. However, bending the card or storing it in a tightly packed wallet with a plethora of other cards, jammed in your pocket is likely to stress it, which may lead to it failing. Whilst I haven't broken a charge point access card yet (and hope I don't due to the cost to replace them!), I have bent and cracked more than one contactless bank card to date, meaning I was unable to touch and pay in shops. Be warned! Unlike the banks, which will issue you a new bank card without charge, it is unlikely that you will be as lucky with the charging companies.

1.6   Things Can Only Get Better...

With a bit of sensible, joined up thinking, things could be much better of course, for everyone. Even with separate scheme providers, the use of contactless bank card technology would allow pure "Pay As You Go" without the need to register for a scheme. This would vastly simplify things for everyone and reduce the capital cost of the schemes card provision. They could even do away with the scheme cards altogether in theory. Chargemaster have just introduced a new charge point product that will accept contactless bank cards and Electromotive are working on the same.

Scheme providers could issue cards that conformed to ITSO (Integrated Transport Smartcard Organisation), or an equivalent standard, as currently being introduced on various public transport networks as smart card ticketing schemes. This is similar to London's Oyster Card, which interestingly being an older technology is not to the ITSO standard. Such joined up access would of course then allow the opportunity to use your ITSO enabled travel card to park at a station or park and ride site and pay for your car to be recharged, before you jump on a train, tram or bus to continue your journey to work, for example. All billing would then be undertaken through your travel card. Simple and the technology exists to do this now, but sadly, it is not being implemented in this manner – yet.

Above: Despite joined up thinking in having an EV charge point at a Park and Ride tram stop, you still need to have separate access arrangements for both - an access card for the charge and ticket for the tram. A Vauxhall Ampera recharges at a GMEV charge point at the East Didsbury Park and Ride site. A Metrolink tram waits at the station behind. **Right:** *A solar panel provides green electricity to an Ampera.*

2.0 EV Benefits

There are a variety of benefits to owning an EV. These include:

1. No road tax.
2. Lower running costs. Source London has estimated that running costs could be up to £1000 cheaper to drive 10,000 miles in an EV than a petrol car.
3. Lower insurance costs. My insurance costs less for my EV than it did for the ICE I had before.
4. Lower maintenance costs. There are less mechanical parts on an EV to wear out and fail.
5. More stable energy prices – electricity versus oil.
6. Zero emissions at the point of use, so no pollution. I won't get into the discussions about how the electricity is generated, but green energy is increasing in the mix of what is used.
7. Better and more efficient use of energy through regeneration during braking
8. Lower carbon footprint per driven mile (ignoring the manufacturing of the car and the batteries, the source of electricity, etc).
9. Opportunity to use "green energy" if you recharge through the likes of Ecotricity or have solar panels at home.
10. You don't pay the London Congestion Charge if driving into central London.
11. Opportunity for free electricity from some charging scheme providers and your own solar panels, if you have them.

Most if not all of the above also apply to an EREV when in battery mode. Some won't apply of course, like zero emission at point of use, if using the ICE on an EREV. I have found the maintenance costs for my Vauxhall Ampera, on a fixed term deal, are cheaper over 3 years than I had been paying for my previous car, a diesel Nissan Qashqai.

3.0 Electric Vehicles

At the time of writing this section there are over 20 different electric vehicle types available in the UK. This figure excludes any variations that may be offered on a particular model of EV. As previously stated, by electric vehicle I am including anything that has a battery that can be recharged from an external fixed electrical supply point. These include:

- Electric Vehicles (EV) – battery and electric motor(s)
- Extended Range Electric Vehicles (EREV) or Range Extender (REx) – battery, electric motor(s) and small Internal Combustion Engine (ICE)
- Hybrid vehicles – ICE, battery and electric motor(s)

I have not included the "original" hybrids such as the Toyota Prius or Honda Insight that cannot have the battery charged by an external fixed electrical power supply. The full list of electric vehicles that are available in the UK as either new or second-hand at the date of publishing is as follows:

3.1 Electric Vehicles

- BMW i3
- Citroen C-Zero
- Ford Focus
- KIA Soul EV
- Mitsubishi i-MiEV
- Nissan e-NV200 Van
- Nissan Leaf
- Peugeot iOn
- Peugeot Partner Van

*Above: A Nissan Leaf EV. **Right:** A BMW i8. **Page 17**: A Tesla Model S on charge*

- Renault Kangoo Z.E.Van
- Renault Twizy
- Renault Zoe
- Smart Fourtwo ED
- Tesla Roadster Electric Car (second hand only now)
- Tesla Model S
- Volkswagen e–Golf
- Volkswagen e–Up!

3.2   Hybrid/Extended Range Electric Vehicles

- Audi A3 e–tron
- BMW i3 REx
- BMW i8
- Chevrolet Volt (second hand only now)
- Mercedes–Benz C–Class Plug in Hybrid
- Mercedes–Benz S–Class PHEV
- Mitsubishi Outlander PHEV
- Porsche Cayenne Hybrid
- Porsche Panamera Hybrid
- Porsche 918 Spyder
- Toyota Prius Plug–In
- Vauxhall Ampera (second hand only now)
- Volkswagen Golf GTE Plug-In
- Volvo V60 Plug–In
- Volvo XC90 Hybrid

4.0 Recharging Costs

Recharging cost information seems to be very difficult to come by for some reason. In fact even schemes that are offering free electricity across all its chargers, such as Ecotricity, GMEV, inmotion! and Tesla*, don't seem to want to shout about it to attract people to electric vehicles, or to join their scheme (were applicable). This seems somewhat counter productive.

Fortunately Charge You Car (CYC) was a little bit more forthcoming with information, although it was buried in an FAQ. Firstly, CYC stated that for them, the tariffs are set by the charge point owner, and not by them. This is fair enough, as the EV charge points are not installed by CYC, but are typically installed by the owner of the car park/premises the charger is located in. This is typical for most of the schemes. Secondly they advise that the recharge costs should be clearly stated on the charge point as well as their own interactive, online map. Thirdly, expected CYC costs are suggested as:

- £3.50 for a fast charge
- £5.00 — £8.00 for a rapid charge

I will explain the difference between fast and rapid charge types later in the book.

It should be noted that some schemes have a minimum charge time of 1 hour stated in their Terms and Conditions. I am not sure how this would work out with a rapid charge, but if you stop for 45 minutes on a fast charge and only recharge for this time, you are likely to be charged for an hour's worth of use. Personally I believe that if EV charging is marketed as "Pay As You Go", this should really mean that you only pay for what you use and no more. I feel that this one hour minimum charge is a bit sneaky and not in the spirit of the desire to promote electric vehicle take up. It would be like petrol stations rounding up to the nearest litre, or nearest £10 of fuel (OK, I know they already round up to the nearest penny), but to the nearest hour seems excessive, with no real justification (except for apparent profit maximisation).

*Tesla factor the electricity cost of their Tesla Super Chargers into the price of the Model S car.

4.1 Electricity Costs

For me, the main reason for having an EV is the environmental driving benefits. This is putting aside how and where the car was made or how the electricity was generated. At home, I use Ecotricity as my electricity (and gas) supplier. This is because they supply electricity that has the lowest number of grams of CO_2 per kWh of any UK energy supplier (8 grams out of a theoretical maximum of 837 grams of CO_2 per kWh based on 100% coal fired electricity generation). You would think that this would come at a premium, but putting aside the daily standing charge (which I pay anyway for the house), I pay 13.33p (£0.1333) per kWh (excluding 5%VAT). So recharging my Ampera from empty to 42 miles of range in 4 hours costs me £1.68 (including VAT at 5%) or just under 4p/mile. But additionally as I have an EV registered with Ecotricity, they give me the equivalent of at least 1,000 "free" miles of electricity each year in the form of a £40 discount from my bill.

4.2   Paying for Charge Scheme Electricity

On all schemes where you have to pay for electricity, you will have to register a debit or credit card to the account and the access card that is associated with it. For Ecotricity, you do not have to register a payment card, as all their charge points are free to use, as well as providing green electricity. But you do have to register to get the access card.

Some organisations take the charge payment pretty much at the time of charge point use, or immediately afterwards. Just like paying for an item in a shop on a credit or debit card. However, some present a bill at the end of the month for all the charging you have undertaken in the month, and then take the money a few days later. This is a bit like a mobile phone bill. At least one scheme requires you to have a credit balance on your account at all times in order to access the charge points in its network. This is on top of an annual fee and a charge point initiation fee.

For most schemes, bills and charge point use can be viewed online, through the same web portal that you used to sign up your card details on originally.

5.0 Car Charging Sockets

The choice of charging connector is typically determined by whether an EV is charged using AC (Alternating Current) or DC (Direct Current), the charging speed (kW power) and the safety protocol employed. In addition, as EVs are designed and manufactured in a variety of countries around the world by a variety of different manufacturers, working to a variety of different standards, the type of connectors used does vary considerably, although often doing exactly the same job.

5.1 AC Car Charger Sockets

AC car charge sockets are detailed in the table below. Note the socket symbols are for information only and are not to scale.

AC Car Charger Sockets	Appearance
Industrial Commando (IEC 60309)	
American Type 1 (SAE J1772) – 5 pin male socket	
European Type 2 (Mennekes, IEC 62196) – 7 pin male socket, though on some EVs the bottom 2 pins may not be present. See photo below	
European Combined Charging System (CCS or 'Combo' 2) – 7 pin male socket, but _only_ using the Type 2 socket (upper part) for AC charging (DC CCS part blanked off)	
Tesla Supercharger connector (European version) – Modified Type 2 (Mennekes, IEC 62196) – 7 pin male keyed socket*	

Above L to R: Type 1, Type 2 and Type 2 CCS (lower DC part blanked off)

5.2 DC Car Charger Sockets

DC car charge sockets are detailed in the table below:

DC Car Charger Sockets	Appearance
Japanese JARI/JEVS G105 or CHAdeMO – 10 pin male socket	
European Combined Charging System (CCS or 'Combo' 2) – 7 pin male socket, but only uses 5 pins for DC charging	
Tesla Supercharger connector (European version) – Modified Type 2 (Mennekes, IEC 62196) – 7 pin male keyed socket*	

The Tesla Supercharger car socket accepts AC or DC inputs, hence why it is shown in the AC and DC tables.

Above L to R: A CHAdeMO and Type 2/CCS DC car charge

In addition to the above, a few electric vehicles have a tethered fixed lead attached to them with a connector plug on the other end, as opposed to a socket. These have the following connector plugs hard wired on:

AC Tethered Plug	Appearance
UK 3–pin plug (BS 1363). The equivalent country specific domestic plug are used by other countries.	
Industrial Commando (IEC 60309)	

6.0 Charging Connector Types

Each EV socket is usually associated with a different charger type and these are shown below. What is disappointing is there is not yet a universal charge point that has all the different sockets and/or cables and the capacity to be able to charge a variety of different EV types from one charger unit. After all a standard petrol pump that offers unleaded and diesel fuels can refill any vehicle regardless of its type.

6.1 AC Fixed Power Supply/Charger Connectors

The various AC fixed power supply/charger connectors are as per the table below:

AC Fixed Power Supply/Charger Connectors	Appearance
UK 3–pin socket (BS 1363). The equivalent country specific domestic sockets are used in other countries	
Industrial Commando (IEC 60309) socket	
American Type 1 (SAE J1772) – 5 pin female connector	
European Type 2 (Mennekes, IEC 62196) – 7 pin female connector socket or tethered socket depending on charge current	

6.2 DC Fixed Power Supply/Charger Connectors

The various DC fixed power supply/charger connectors are as per the table below and all are on the end of a tethered cable. There are no DC sockets.

DC Fixed Power Supply/Charger Connectors	Appearance
Japanese JARI/JEVS G105 or CHAdeMO – 10 pin female connector tethered	
European Combined Charging System (CCS or 'Combo' 2) – 5 pin female connector tethered	
Tesla supercharger connector – Type 2 (Mennekes, IEC 62196) – 7 pin female connector tethered	

7.0   Charge Point Types

As in the early days of video tape players and recorders, when both VHS and Betamax were vying to become the industry standard, electric vehicles unfortunately have a variety of different charge connection types. Some of this is down to the development of the technology, some down to personal manufacturers design philosophy, some to different standards and some of it would seem to be down to a general disregard for interoperability and non industry co–operation, in what still remains a very niche market in terms of world wide automotive sales.

In addition to the different charge sockets, you also need to be aware of the speed of the charge. Vehicles that can be recharged faster need to have higher rated chargers and charge cables, that are either higher voltage, higher current or DC rather than AC electricity.

The different charger types are explained in the next sections.

7.1   Slow Charge Points

This is basically a domestic 230V AC 13A 3 pin socket as you would have in your house which supplies power at up to 3kW (13A) to your electric vehicle. Typically a full charge for an empty battery would take up to 8 hours. Most electric vehicles have the option to plug in to a domestic supply, via a special cable (3 pin plug at one end and the specific female vehicle connector at the other end) which may be included with the vehicle or as an optional extra, which does somewhat amaze me, as there are 3 pins sockets all over the country. This is great if you go somewhere that doesn't have charging infrastructure and are staying the day or staying over night, so allowing you to get a full charge. Some of the early charge points were fitted with a 3 pin socket to allow this type of charging.

Right: A domestic 3 pin socket charge point. The switch is located above.

7.2 Fast Charge Points

These are charging points that use either "domestic" single phase (up to 7kW) or three–phase 230V AC electricity (up to 22kW) all with a supply rating of between 16 and 32A supply. As these units are higher rated, they have industrial power sockets on the charge units so you cannot plug in anything apart from an electric vehicle. The charge point is often known as a "Type 2" or "Mennekes" due to the connector socket fitted. Mennekes is the brand name of the manufacturer of the socket/connector, Mennekes Electric Ltd, a leading German manufacturer of industrial plugs and sockets. Despite being referred to as a fast charger, the time taken to recharge a typical EV can be up to 4 or more hours. They come in a variety of different shapes and sizes:

A Guide to UK and Ireland Electric Vehicle Charge Points

Below is a list of UK electric vehicles that can use the Fast charge points and the associated cable they require to connect to them (required vehicle connector to charge point connector):

7.2.1 Fast Type 1 Vehicle Connectors

Below is the list of EVs and EREVs that have a Type 1 vehicle charge post socket and the cable type they need to connect to a Type 2 Fast charge point.

EV Cable Connector	Cable	Charge Point Connector
Type 1 Female	3, 5 or 10m cable	Type 2 Male

Vehicle Type	Vehicle Model
Electric Vehicle	- Citroen C–Zero - Ford Focus - KIA Soul EV - Mitsubishi i–MiEV - Nissan e–NV200 Van - Nissan Leaf - Peugeot iOn - Peugeot Partner Van - Renault Kangoo Z.E
Hybrid/Extended Range Electric Vehicles	- Chevrolet Volt - Mitsubishi Outlander - Toyota Prius Plug–In - Vauxhall Ampera

7.2.2 Fast Type 2 Vehicle Connectors

Below is the list of EVs and EREVs that have a Type 2 vehicle charge post socket and the cable type they need to connect to a Type 2 Fast charge point.

EV Cable Connector	Cable	Charge Point Connector
Type 2 Female	3, 5 or 10m cable	Type 2 Male

Vehicle Type	Vehicle Model
Electric Vehicle	• BMW i3 EV • Renault Zoe • Smart Fourtwo ED • Tesla Model S • Volkswagen e–Golf • Volkswagen e–Up!
Hybrid/Extended Range Electric Vehicles	• Audi A3 Sportback e–tron • BMW i3 REx • BMW i8 • Mercedes–Benz C–Class • Mercedes–Benz S–Class • Porsche Cayenne • Porsche Panamera • Porsche 918 Spyder • Volkswagen Golf GTE Plug-In • Volvo V60 Plug–In • Volvo XC90

7.3 Rapid Charge Points

There are two different types of rapid charger in existence – AC and DC (Direct Current) versions. The charge points are able to supply up to 43kW of AC power or up to 50kW of DC power and are able to recharge a nearly depleted battery back to 80% charge in around 30 to 60 minutes, depending on the model of EV. Due to the power these charger units are able to deliver, all are provided with a tethered charge cable complete with a non–removable vehicle connector on the end. In addition, due to the charging infrastructure currently installed, it is often only possible to rapid charge one vehicle at a time as the power required is just too great to recharge any more. However some newer model charge point units allow two or more EVs to simultaneously charge. I will discuss each charger type in turn below.

Page 23: A Peugeot iOn car and Peugeot Partner van.

7.3.1 Rapid AC Charge Points

Rapid AC chargers are typically rated at 43kW (3 phase, 63A) and have a Type 2 Mennekes connector on the end of a tethered cable. They are relatively new development in the UK and are only currently available to a few EV models to provide a rapid AC charge. Below is a list of UK electric vehicles that can use the Rapid AC charge points:

- Renault Zoe
- Smart Fourtwo ED (only with the optional 22kW on–board charger fitted to the vehicle)

Additionally, an AC rapid charge point will also Fast charge other EVs with a Type 2 socket at the lower Fast Charge rate, via the same tethered cable connector, as previously described under Fast Charge Points (see 7.2.2). The other EVs are listed below:

7.3.1.1 Electric Vehicles that will Fast Charge on a Rapid Charge Point

- Audi A3 Sportback e–tron   Type 2
- BMW i3   Type 2/CCS Combined*
- Tesla Model S   Type 2
- Volkswagen e–Golf   Type 2/CCS Combined*
- Volkswagen e–Up!   Type 2/CCS Combined*

7.3.1.2 Hybrid/Extended Range Electric Vehicles that will Fast Charge on a Rapid Charge Point

- BMW i3 Rex   Type 2/CCS Combined*
- BMW i8   Type 2
- Mercedes–Benz C–Class   Type 2
- Mercedes–Benz S–Class   Type 2
- Porsche Cayenne   Type 2
- Porsche Panamera   Type 2
- Porsche 918 Spyder   Type 2
- Volkswagen Golf GTE Plug-In   Type 2
- Volvo V60   Type 2
- Volvo XC90   Type 2

*Using the Type 2 connector part only and not the CCS DC connector. See section 5.1 AC Car Charging Sockets for details

Above: *An Ecotricity branded DBT-CEV rapid charge point with the AC tethered charger cable. This is at Welcome Break Keele services, on the M6 northbound. A Type 2 fast charger can be seen in the background.*

7.3.2 Rapid DC Chargers

Rapid DC chargers provide a high power DC supply at the connector with power ratings of up to 50kW (125A). Charge units are fitted with a tethered cable with a non-removable JEVS (CHAdeMO) or Type 2/CCS (Combo) vehicle connector. These will typically rapid charge a depleted battery to 80% capacity in around 30 minutes.

Below is a list of UK electric vehicles that can use the Rapid DC charge points:

- BMW i3   CCS (if fitted - optional extra)
- Citroen C–Zero   CHAdeMO
- Mitsubishi I–MiEV   CHAdeMO
- Mitsubishi Outlander PHEV   CHAdeMO
- Nissan LEAF   CHAdeMO
- Nissan e–NV200   CHAdeMO
- Peugeot Ion   CHAdeMO
- Peugeot Partner Electric Van   CHAdeMO
- Volkswagen eUp!   CCS
- Volkswagen eGolf   CCS

The DC rapid chargers will charge an EV battery *faster* than an AC rapid charger.

7.3.3 Why are there two versions of Rapid Charger?

When you analyse a typical set up of an EV and its associated charging system, it contains a number of differing technology types. These include:

- AC charge point – which charges a...
- DC Battery – which powers the...
- AC motor – which turns the wheels and regenerates back to the DC battery

To change AC to DC or DC to AC a conversion process is required. AC to DC is converted via a rectifier. DC to AC via an inverter. These conversions are not 100% efficient as losses occur in the form of sound and some heat. Of course the reason for the conversions is that an AC battery does not exist and an AC motor offers many benefits including size (smaller), power output (higher), maintenance (less) and weight (lower) when compared to a DC motor.

In addition, with the need to be able to charge EVs faster, there is a requirement to put more power into them. However, larger amounts of high

Above: *An Ecotricity branded DBT-CEV rapid charge point with the DC CHAdeMO tethered charger cable. This is the other side of the same charger on page 26 and is an earlier example of a rapid charge point with just two connections types.*

current AC power being pushed in from the charge point to the EV means a larger, more powerful rectifier is required in the EV to convert it into DC power for the battery. Above 75A it is typically better for the charge point to supply the DC power to the battery, as the cost, size/weight and thermal issues of the conversion limit how much power a suitably sized rectifier that will still fit into the EV can handle. This is the reason why the rectifier is fitted into the DC rapid charging unit and why it is much larger than a standard AC fast charger. In fact a rapid charger is often the size of a petrol pump! DC rapid chargers push large amounts of DC power straight into the EV battery and this is why they are faster to recharge a depleted battery than the AC rapid chargers.

Rapid charging is all well and good but of course comes at a price of battery life. If you can avoid rapid charging and stick to fast charging or slower, then you will get better longevity out of your battery. Vauxhall will not allow rapid charging on the Ampera for this very reason, even though in normal driving conditions the battery is frequently rapid charged when using regenerative braking. Mitsubishi is the only manufacturer I have seen that openly encourages occasional rapid charges.

7.3.4 Tesla Supercharger

Tesla Motors have gone one better than the current standard DC rapid charger. They have taken the concept and upped the power output to 120kW (160A). This means a Tesla Model S, the only car that can currently be connected to a Tesla supercharger, can be recharged in mere minutes, as opposed to hours. The Tesla supercharger uses a standard Type 2 female connector, but will only communicate with a Model S to deliver the supercharge.

Right: *A Tesla Supercharger in Warrington*

29

Above: An inmotion! Branded Siemens QC45 rapid tri-charger charge point at the Meadowhall shopping centre in Sheffield. The three connectors are (L to R) CHAdeMO, CCS and Type 2.

7.3.5 Tri–charger or Triplex Rapid Charger

You may see the terms tri–charger or triplex charger banded about. This is basically a single rapid charge unit fitted with three tethered cables. These provide a rapid Type 2 AC connection, a rapid DC CHAdeMO connection and a rapid DC Type 2/CCS connection. The term is used to distinguish it from the earlier rapid chargers that were only fitted with two tethered cables and so only two different types of connector (usually a rapid Type 2 AC connection and a rapid DC CHAdeMO connection).

7.3.6 UltraCharger

This is not another faster charger type, but a UK designed and built rapid charger which I have included in this section to explain and avoid confusion. Chargemaster, the company behind the Polar network, unveiled the UltraCharger on 9 September at the Low Carbon Vehicles 2015 event at Millbrook Proving Ground, Bedfordshire.

The UltraCharger is considerably smaller than the current range of rapid chargers installed, so planning requirements are less onerous when considering installing one. It is also a more aesthetically pleasing unit and somewhat less industrial looking. The charge cables neatly retract under the front of the unit, so keeping the footprint small, as well as keeping the cables safely stored out of the way – clean, safe and dry. The charger will accept contactless bank cards and is fitted with a large colour touchscreen, which can be programmed to display a personalised welcome message when used with a scheme RFID card. The unit is fitted with rapid Type 2 AC, rapid DC CHAdeMO and rapid DC Type 2/CCS, the same as a tri–charger or triplex charger.

7.4 Charge Point Back Office Standards

Despite all this fragmentation I have mentioned so far in terms of both technical designs and implication of charging networks, there is some joined up thinking, in the form or an emerging global industry standard called the Open Charge Point Protocol (OCPP – not to be confused with POD Point's Open Charge Smartphone App and charge points). This is an open protocol that is becoming the most widely adopted and used standard for the connection of charge point equipment to the back office management systems, used for monitoring and billing. Charger manufacturers such as APT Technologies, Chargemaster, DBT-CEV, Elektromotive and efacec have already adopted this protocol in their equipment.

Below: *An UltraCharger 500R seen on the Chargemaster stand of the Low Carbon Vehicles 2015 event at Millbrook Proving Ground in Bedfordshire. The three connectors are (L to R) CCS, Type 2 and CHAdeMO.*

7.5 Charging Modes

You may hear or read about charger types referred to as Mode 2 etc. This is just a way to categorise the different types of charger and is part of the IEC 62196–1 Standard. In simplistic terms it means:

Mode	Details
1	Slow domestic AC Charging using a 3 pin plug. This is not used in the UK.
2	Slow domestic AC Charging via the built in control box (the plastic box fitted on the charge cable) and using a 3 pin plug. This is the norm for all EVs.
3	Fast or Rapid AC Charging from a charge point. This would include your professionally installed fast charge point at home or a charge point in the street or at the motorway services etc
4	Rapid DC Charging from a charge point.

8.0 Joining an Electric Vehicle Charging Scheme

For most of the schemes listed in this book, the following typically applies:

1. You need to register with the charger scheme provider. A one off or annual fee may be payable or it may be free. Details vary between provider.
2. You may have to provide card payment details or bank account details to the provider, even if you never intend to use a charge point where you will be required to pay for a charge.
3. You will receive a charge point card access through the post if you have opted for this. This is usually a contactless RFID access card, so take care not to bend or crack the card as it will then not work.
4. Drive to a charge point that is suitable for your vehicle. You can find details online, on the charge companies websites.
5. Follow the instructions on the charge point to use it. Use the RFID access card to activate it in order to start charging your vehicle. Or if you don't have an access card and have opted to use a mobile Smartphone App, then follow the instructions to use this.
6. Charges may apply for electricity used. **Note that you may also be charged a minimum of one hours use, even if you only charge for (say) 30 minutes**. Check with the scheme provider for exact details.
7. **Car parking fees may also apply where the charge point is located**. Check with the car park location so you don't come back to find a parking fine slapped on your windscreen, or even worse a wheel clamp.
8. When you return to your vehicle, you usually have to present your card to the charge point again in order to stop the charge and unplug from your vehicle. Again, follow the instructions on the charge point to use it, as there are differences.

In addition to the above, on some schemes it is possible to call a dedicated number to access the charge point. The telephone number is printed on the charge point in a similar manner to calling to pay for car parking at certain locations. Personally I find this a lot of hassle and the access card is far easier.

The new Polar UltraCharger, where installed, will allow contactless bank card access, which will make the need for a membership a thing of the past.

8.1 Charge Point Location Maps

There are a variety of websites that offer the facility to locate electric vehicle charging points. These can roughly be split into those sites that are only promoting their own scheme and those that are promoting all schemes. What I have found from experience is:

1. There are anomalies in the data provided, for instance charge points that have been removed. I often check two different websites to verify a site actually exists, as I have found one site to incorrectly show the removed charge point *is present*, and another showing it has been removed!
2. The location addresses provided are often difficult to find in reality, especially if you are desperate for a charge! This is not really the fault of the mapping websites, as they are only reproducing the data they have been provided with. Additionally, lack of signage or poor signage at the charge point location does not help alleviate the situation either. If you are planning a trip, and you know you may end up being desperate for a charge, it does pay to check the likes of Google Street View to help locate where exactly you need to go, although the charge points, being relatively new, are rarely featured on them at the moment. However, it will still give you a better idea of the general area you are driving to.
3. Locations are listed where there is restricted access/no public access, but this may not always be that obvious. You do need to check the "small print" for the access details. I have visited one site at 1030 on a Sunday morning where restricted access was described as "Access restricted to council employees between 08:00 – 18:00 Mon to Sat. Free outside these hours." This implied that it was pen to the public outside these hours. However there were two Nissan Leaf's and a Vauxhall Ampera parked up at the charge point when I visited, but none were charging!

Mapping websites include:

- www.ChargeMap.com
- www.OpenChargeMap.org/app/
- www.PlugShare.com
- www.Zap–Map.com/location–search

Above: Spot the missing GMEV (CYC) charge point at Travis Street in Manchester

9.0 Charge Point Use

When using a charge point, there are a few things you should be aware of. These are listed below.

9.1 Using Your Own Charge Cable

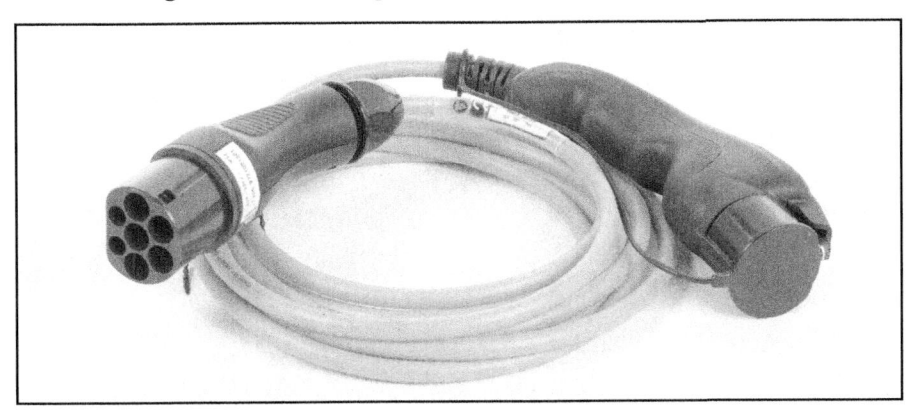

- If your EV has the ability to delay the charging start, make sure it is set for an immediate start. Unless you want a delayed start time to charging
- Make sure your cable and connectors are dry and undamaged before using. Also make sure your hands are dry and they do not come into contact with the terminals on the car socket or charge connector
- Fully unwind your cable before use.
- Never use an extension lead to charge your vehicle.
- If using the domestic 3 pin charge cable connector, make sure the box of electronics cannot get wet
- Connect your car first before connecting to the charge point.
- Connect the charging cable at both ends before beginning the charge.
- Make sure the charge cable is not left trailing so creating a trip hazard by your car. If you can, tuck the cable under your car, out of the way.
- Make sure you lock your car when leaving it on charge.
- End the charging process before removing the cable from your vehicle. Best to disconnect from the charge point first then from your car.
- Charging cables trailing on the ground will inevitably get wet and dirty. It is a good idea to have a pair of gloves and a cloth with you. This is to keep your hands clean and to wipe the cable down before putting it away, especially if it has been raining. A supermarket "bag for life" makes an ideal cheap but durable bag to store your cable in, if you didn't get one with your cable/vehicle. Alternatively you can buy a cable bag from the likes of www.EVConnectors.com (other companies may also offer similar).

Below: *If you didn't get a storage bag for your charge cable, then EV Connectors make this useful product. As well as a zip that opens half way round the circumference for easy access and carry handles, it has "Velcro" hooks on the back to securely fasten in your car. I was very excited when I discovered and purchased this product!*

- If one of the connectors has a cover cap with it, put this back on the connector end when not in use. This is to keep "foreign objects" from entering into the connector and causing you charging problems or car/cable socket damage.
- When you've finished charging, recoil your cable in the manner it wants to coil and store it somewhere safe, and dry, protecting it from the risk of accidental damage. Try and avoid coiling the cable against its natural "coil", as this may cause damage to the cable.

9.2 Using Your Own Charge Cable at a Public Charge Point

When using a charge point with your own cable, please note the following:
- Only park in a public charge point space if you intend to charge your EV. It is not privileged parking.
- If your EV has the ability to delay the charging start, make sure it is set for an immediate start. Otherwise you will be parked at a charge point and not be recharging!.
- Make sure your cable and connectors are dry and undamaged before using. Also make sure your hands are dry and they do not come into contact with the terminals on the car socket or charge connector
- Fully unwind your cable before use.

- Follow the instructions on the charge point exactly as detailed. Even if you are familiar with using charging points, follow the instructions, as there can be subtle differences.
- Connect your car first before connecting to the charge point.
- Connect the charging cable at both ends before beginning the charge.
- Make sure the charge cable is not left trailing so creating a trip hazard by your car. If you can, tuck the cable under your car, out of the way.
- Make sure you lock your car when leaving it on charge.
- End the charging process before removing the cable from your vehicle. Best to disconnect from the charge point first then from your car.
- Charging cables trailing on the ground will inevitably get wet and dirty. It is a good idea to have a pair of gloves and a cloth with you. This is to keep your hands clean and to wipe the cable down before putting it away, especially if it has been raining.
- If one of the connectors has a cover cap with it, put this back on the connector end when not in use. This is to keep "foreign objects" from entering into the connector and causing you charging problems or car/cable socket damage.
- When you've finished charging, recoil your cable in the manner it wants to coil and store it somewhere safe, and dry, protecting it from the risk of accidental damage. Try and avoid coiling the cable against its natural "coil", as this may cause damage to the cable.
- Report any damage or other charge point anomalies to the charge point scheme provider, so action can be taken as required, to keep the charge point fit for use for all. You will appreciate others doing this when you need a charge.
- If possible, when you have finished charging, move your car away, so allowing someone else to use the charging point.

9.3 Using a Tethered Charge Cable on a Rapid Charge Point

When using a rapid charge point with a tethered cable, please note the following:

- Only park in a rapid charge point space if you intend to charge your EV. It is not privileged parking.
- If your EV has the ability to delay the charging start, make sure it is set for an immediate start.
- Make sure the rapid charge point cable and connector has no visible damage before using it.
- Fully unwind/unfurl the cable before use.

- Follow the instructions on the charge point exactly as detailed. Even if you are familiar with using rapid charging points, follow the instructions, as there are differences.
- Connect the charge point connector to your vehicle when instructed.
- Make sure the charge cable is not left trailing so creating a trip hazard by

Above: A good example of how you shouldn't leave a charge point. The second and third cables are trailing on the floor and should be hung up out of the way, as they are tripping hazards. The third connector (Type 2 AC) is not even inserted into the storage holster, but lying on the ground.

your car. If you can, tuck the cable under your car, out of the way.
- End the charging process before disconnecting the cable from your vehicle.
- Charging cables trailing on the ground will inevitably get wet and dirty. It is a good idea to have with you a pair of gloves and a cloth. This is to keep your hands clean.
- When you've disconnected the charger cable from your vehicle, put the connector back in the "holster" on or by the rapid charge unit, making sure the cable doesn't cause a trip hazard where it is lying.
- Report any damage or other charge point anomalies to the charge point scheme provider so action can be taken as required to keep the charge point in use for others. You will appreciate others doing this when you are desperate for a charge.
- When you have finished charging, move your car so allowing someone else to use the rapid charging point.

9.4 Rapid Chargers and Heart Conditions

The following safety warning was seen posted on a SSE installed APT Technologies "Tri–Rapid Compact charger" and is quoted verbatim:

"People who use electronic medical devices such as implanted cardiac pacemakers or implantable cardioverter–defibrillator (ICD) might be affected by electric wave from the quick charger. Keep the device away from the quick charger more than 2 meters while is in charge operation."

So according to this notice, if you have either a pacemaker or implanted defibrillator then you cannot use a rapid charge point, as you will need to go within 2m of the unit to start and stop it. Somewhat worrying really!

9.5 ICED

When a standard Internal Combustion Engine (ICE) vehicle parks in an electric vehicle charge point bay, so blocking it for EV use, the charger is said to have been "ICED". This is generally illegal under the terms of the installation of the electric vehicle charge point. However, it is not typically enforceable by the police if the EV charge point is on private land, including council land. The car park owner does have the right to issue a parking ticket to the vehicle owner, if the car parking owner is aware of the issue, and/or in some instances tow the offending vehicle away.

Like most car parks, some are heavily patrolled, with an attendant on site in a cabin and/or patrolling around. Other car parks have no more than a sign on a wall with lots of small print, which no one ever reads, let alone understands. So if a charge point is ICED you should report it to the car park owner/operator who may be:

- Public car park: Operating company, via the on site representative, if available
- Shop or business car park: shop or company facility manage etc
- Contact the charge point company and report it to them, at the time of witnessing the ICED vehicle
- Contact the local authority for the area and report it to them, especially if it is a local council car park or on street parking

You can also take a picture of the offending vehicle, making sure that you have the registration and the charge point visible in the pictures so it is obviously parked in an EV charge bay. You can provide this to the charge point company/car park owner as evidence. Make a note of the date and time as well. Some people have been known to post pictures of offending vehicles on social media/websites, such as "Spotted Parking Like a Twat" (SPLAT), or on the charge companies page. However, I would never suggest you do something like this...

If you are really desperate for a charge, then in some instances it may be possible to access the charge point whilst not actually being parked in the charge point bay for the charge point. I managed do this once at a Type 2 charger, where the actual charge bay was ICED but I could park in the next

Above: *A funny looking EV is parked at a Polar charge point at a Sainsbury's in Bath*

Below: At the Milton Keynes Coachway, a Polar operated tri-charger has three ICE vehicles parked in front of it. Being close to the M1, this is not a location an EV driver wants to find ICEd, but the lack of signage or road marking do not help the situation.

bay along. My cable was just long enough to reach over my bonnet, past the offending ICE and to the charge point. Not an ideal situation, but if needs must, it is worth a go. Alternatively, by ringing the charge point operator (or consulting on line mapping if you have it on your Smartphone) you may be able to find another charge point locally that you can use and more importantly get to. But as charge point infrastructure is still a bit thin on the ground in a lot of places, I wouldn't hold my breath...

Additionally and rather selfishly, I have seen charge points blocked by EVs that are parked but not even connected to the charge point! Remember, charge points are not parking spaces for EVs.

Above: Confusing road markings and signs don't help charge points stay ICE free.

10.0 Charge Point Scheme Providers

Charge point schemes are typically regional in nature, apart from Ecotricity, which has created what it calls the "Electricity Highway". This is a nationwide network, which has 229 rapid and 21 fast charge points, installed initially at UK motorway services, but has since been expanded to other locations including IKEA stores, for example.

All the regional schemes are usually run by an organisation that nationally operate a number of different schemes. In some instances a card issued on one scheme can be used on a different scheme if it is run by the same operator. Ownership and operation of charging infrastructure should not be confused. From a driver's perspective most schemes do not own the infrastructure you may use, so car parking restrictions including car parking fees may apply, and/or there may be electricity costs.

The regional schemes are:

Scheme Name	Geographical Area	Provider
Chargernet	Dorset	Charge Your Car
Electric Corby	Corby	Polar
Energise	South East	Charge Your Car
GMEV	Greater Manchester	Charge Your Car
Greener Scotland/ ChargePlace Scotland	Scotland	Charge Your Car/ Ecotricity
Hackney	Hackney	Source London/ ChargePoint Genie
inmotion!	South Yorkshire	npower Electric Vehicles
Milton Keynes Crosslink	Milton Keynes and surrounding area	Polar
Plugged–in–Midlands	Midlands	Polar
Recharge	Merseyside/West Cheshire	Charge Your Car
Source East	East Anglia	POD Point
Source London	London	Source London
Source West	South West	Charge Your Car
Think Travel/Think Electric	Gloucestershire	Charge Your Car

The nationwide schemes or organisations that cover a variety of areas or an entire country are:

- ChargePoint Genie
- Charge Your Car
- ecarNI (Northern Ireland)
- Ecotricity
- ESB (Ireland)
- Just Park
- POD Point
- Polar
- SSE
- Tesla Supercharger Network
- Zero Carbon World

What you do need to be aware of though is that even though a particular scheme may be dominant in a particular area, you are still likely to find other organisations providing charge points in the same area. For example in the Plugged–In–Midlands area, now operated by Polar, you will also find Charge Your Car and POD Point charging points, meaning you will likely need a multitude of access cards.

I will run through all the schemes in alphabetical order.

Above: An inmotion! tri-rapid and two Type 2 charge points at Meadowhall, Sheffield.

10.1 ChargePoint Genie

Geographical Area: Hampshire, Isle of White, Cornwall, Oxfordshire (1x) and Hackney

- Website: www.ChargePointGenie.com
- Charger Types: Type 2 Fast and Rapid AC/DC Chargers
- Operated by: ChargePoint Genie
- Access: Genie Smart Card, or Genie Smartphone App
- Access Cost: £20 including VAT per Annum.
- Electricity Charges: See below

- Telephone: 020 3598 4087 – 24 hours a day
- Email: GenieSupport@ChargePointServices.com

10.1.1 Details

ChargePoint Genie provides the smart card access for SSE installed chargers in Hampshire (including the Isle of White), Cornwall, and one charger at Eynsham, in Oxfordshire. They also provide the same provide the back office access to three rapid chargers in Hackney, London. Frustratingly, unlike other scheme operators, an SSE Genie Smart Card won't allow you access to the Hackney chargers. Likewise a Hackney Genie Smart Card won't allow you access to the SSE chargers!

Like most schemes, there is an RFID access card charged at £20 a year as well as a Smartphone App. Electricity is charged at £0.30 per kilowatt/hour (kWh) used. You have to keep a credit balance in your account in order to use the charge points and there is a minimum top of purchase of £1.00. So, not strictly speaking Pay As You Go unlike some other schemes.

ChargePoint Genie also has some other interesting differences including:

- Charge Initiation fee. You will pay £0.60 on a fast charge and £1.80 on a rapid charge, just for commencing the charge.
- Overstay charge. You will pay £10.00 if your vehicle is left on charge for more than 4 hours (fast charge) or after an hour on rapid charge. For a rapid charge this is repeated every hour.
- ChargePoint Genie have restricted the rapid charger to only deliver an 80% charge and then cut out. If you want to get that 20% extra so you have a full battery, you need to start the charge again, which will mean another

initiation feed of £1.80!
- As previously stated, unlike other operators, a ChargePoint Genie Smart Card will not access all the charge points that it runs the back office for.

The overstay charge has been levied to make sure that there is a turnover of people using the charge point and it is not hogged, especially the rapid. The idea being that a driver who needs a rapid charge will not have to wait more than an hour to speak to the driver of the vehicle on the charge point.

Based on the prices stated, a 30 minute charge on a DC 50kWh rapid charger, would cost you:

50kWh / 2 (30 minutes) = 25
x £0.30/kWh = £7.50
+ £1.80 connection fee
= £9.30

For a 30 minute charge on an AC 43kWh rapid charger, would cost you:

43kWh / 2 (30 minutes) = 21.5
x £0.30/kWh = £6.45
+ £1.80 connection fee
= £8.25

For a 4 hour 7kWh fast charge you would expect to pay:

7KWh x 4 hours = 28
x £0.30/kWh = £8.40
+ £0.60 connection fee
= £9.00

For my Vauxhall Ampera, which pulls 3kWh when charging and takes 4 hours on a fast charger from empty, I would expect to pay:

3kwH x 4 hours = 12
x £0.30/kWh = £3.60
+ £0.60 connection fee
= £4.20

This would give me around 42 miles of range depending on the ambient temperature at the time of charging.

10.2 ChargerNet

Geographical Area: Dorset

- Website: www.Poole.gov.uk
- Charger Type: Rapid AC/DC Chargers
- Operated by: Charge Your Car (CYC)
- Access: CYC Charge Your Car card, Charge Your Car Smartphone App, or Pay As You Go via the telephone – 0191 260 5625 – 24 hours a day
- Access Cost: Free or £20 including VAT per Annum.
- Electricity Charges: £4/hour

- Telephone: 0191 265 0500 - office hours 0800 to 1800. Outside office hours for emergency assistance only
- Email: None given

10.2.1 Details

ChargerNet is the name for the rapid EV charging points installed in Dorset under the auspices of Bournemouth Borough Council, Borough of Poole and Dorset County Council. This network was established following the OLEV grant for £900,000 jointly awarded to the three councils. The 18 rapid chargers have been installed across the region close to the main routes between the M27 and Exeter as well as Yeovil and the South Coast. 5 will be in Poole, 5 in Bournemouth, and 8 across Dorset, including Bridport, Christchurch, Dorchester, Lyme Regis, Weymouth and Wimborne.

The scheme is operated by CYC and there is no dedicated ChargerNet access card. You need to apply for a CYC card at www.chargeyourcar.org.uk/ev-driver/join/

10.3 Charge Your Car

Geographical Area: Various schemes across the UK

- Website: www.ChargeYourCar.org.uk
- Charger Types: Type 2 Fast and Rapid AC/DC Chargers
- Operated by: Charge Your Car (CYC)
- Access: Charge Your Car Access Card, Charge Your Car Smartphone App, Pay As You Go via the telephone 0191 260 5625 – 24 hours a day
- Access Cost: Free or £20 including VAT per Annum. See below
- Electricity Charges: See individual schemes

- Telephone: 0191 265 0500 - office hours 0800 to 1800. Outside office hours for emergency assistance only
- Email: admin@chargeyourcar.org.uk

10.3.1 Details

Launched in 2010, Charge Your Car (CYC) operate a number of schemes on behalf of others with the aim of creating the UK's first national EV charging network. From its beginnings in the North East, the network has now expanded nationwide, with over 2000 charging points connected and more being added all the time. However, it does not own any of the roadside charge points. These are owned by the car park/business owner. CYC merely provides the connectivity behind the scenes (the "back office"), providing:

- CYC Access Card
- CYC Smartphone App
- Website with interactive charge map, showing real time charge point status
- Telephone Helpline
- Charge point monitoring, maintenance management and billing platform for charge point owners
- One brand promotion

CYC manages charge points for a variety of public and private sector organisations, including:

- Local authorities – GMEV, Energise etc
- Train operating companies – Southern Railways
- Commercial networks

- Fleets
- Vehicle manufacturers
- Car park operators – NCP
- Supermarkets
- Hotels and B&Bs.

The Charge Your Car Smartphone App allows you to search for charge points, plan a route to a charge point as well as start and end a charge session. A word of warning though. Using the Smartphone App is only as good as your mobile data connection, as some charge points may be located in areas of poor coverage, you are advised not to rely on this alone. In addition CYC advise that some of the charge points may not support access via a mobile telephone. So having a card as well is a prudent idea, but will cost you at least £20/year for the benefit. You are also able to save frequent, favourite or most recently used charge points, and pay for recharging via the CYC Smartphone App.

If you don't want to buy a card, or don't have a Smartphone, you can also use a charge point via your mobile telephone, by calling the 24 hour automated telephone number shown on the charge point. This is a voice recognition system and you will need your credit/debit card to pay for the charge.

10.3.2 CYC Access Card Costs
- You can apply for a CYC card at www.chargeyourcar.org.uk/ev–driver/join/
- Initial cost £20 – but see Top Tip! below
- Automatic renewal £20/year
- Manual renewal £27.50/year
- Electricity costs extra, as used, or free on some schemes

10.3.3 CYC App
- No initial or ongoing fees
- Electricity costs extra, as used, or free on some schemes

10.3.4 Telephone Use
- No initial or ongoing fees, apart from the telephone call costs
- Electricity costs extra, as used, or free on some schemes

10.3.5 Top Tip!
Get a GMEV Access Card for a one off £10 fee. You can use it at all CYC sites. However, GMEV are now talking about restricting this offer to postcodes within the Greater Manchester area only.

10.4 ecarNI

Geographical Area: Northern Ireland and roaming into the Republic of Ireland

- Website: www.ecarNI.com
- Charger Type: Type 2 Fast and Rapid AC/DC Chargers
- Operated by: ESB
- Access: ecarNI Access Card or Smartphone App. ESB ecars Access Card
- Access Cost: Currently Free
- Electricity Charges: Currently Free

- Telephone: 0845 601 8303
- Email: ecarni@esb.ie

10.4.1 Details

Since 2010, ecarNI has installed of over fast 334 electric vehicle charging points at 174 different locations across Northern Ireland. More recently, 14 rapid charge points have been installed, on or near strategic road routes. The intention is that EV drivers are no more than 30 miles from a rapid charger.

ecarNI was developed by a consortium of public and private organisations, including the 20 local councils, the electricity companies of Northern Ireland and the Republic of Ireland as well as Intel and SAP. On the 30th July 2015, the operation, maintenance and development of the ecarNI charge point network was transferred to ESB, who run the Republic of Ireland charge point network. This creates a combined Irish network across island.

10.4.2 Roaming

An ecarNI access card will also work in the Republic of Ireland on the ESB ecars charger network. An ESB ecars access card will also work on the ecarNI scheme now.

Page 46: A BMW i3 Range Extender (REx)

10.5 Ecotricity

Geographical Area: Nationwide but mainly various Motorway Services and Ikea stores

• Website:	www.Ecotricity.co.uk/for-the-road
• Charger Type:	Type 2 Fast and Rapid AC/DC Chargers
• Operated by:	Ecotricity
• Access:	Ecotricity Electric Highway Card. Also see details below
• Access Cost:	Currently Free
• Electricity Charges:	Currently Free
• Telephone:	08000 302 302
	0830 to 1730 Monday to Thursday and 1630 on Friday
	01453 761482
	0800 to 2000 Monday to Friday and 0900 to 1700 Saturday
• Email:	home@Ecotricity.co.uk

10.5.1 Details

In July 2011, Ecotricity installed their first electric vehicle charge point at the Welcome Break services at South Mimms on the A1/M25. This marked the beginning of what they call the "Electric Highway" – the UK's first national motorway network of charge points for electric cars. Interestingly some of the early charge points included 3 pin plug sockets! Ecotricity charge points can now be found at most motorway service stations across England as well as Ikea stores too.

At certain locations where an Ecotricity charge point unit is geographically in a Source London, Plugged–in Midlands, Source East or Charge Your Car area, access cards from these schemes can also be used to recharge. These charge points have signage showing this on the charge point unit. However these are not reciprocal agreements at the moment.

The Ecotricity Electric Highway card is a must have card for all EV drivers if you travel outside of your local area.

Right: A Nissan Leaf uses the CHAdeMO connection on a DBT-CEV tri-charger rapid charge point to charge for free at the IKEA store in Warrington.

A Guide to UK and Ireland Electric Vehicle Charge Points

10.6 Electricity Corby

Corby Area only, although part of the Plugged–In–Midlands scheme

• Websites:	www.ElectricCorby.co.uk
	www.ChargeMasterPLC.com
	www.PolarInstant.com
• Charger Type:	Type 2 Fast and Rapid AC/DC Chargers
• Operated by:	Polar
• Access:	Polar Access Card or Polar Instant Smartphone App
• Access Costs:	Free, £12/month or £20/year. See Polar
• Electricity Charges:	Varies depending on the access scheme. See Polar
• Telephone:	0845 838 0551 – 24 hours a day
	01582 400331 – 0830 to 1730 Monday to Friday
• Email:	PiM@ChargeMasterPLC.com

10.6.1 Details

Electric Corby is a not–for–profit Community Interest Company formed with the backing of Corby Borough Council. It has a number of aims including making Corby:

• A leading edge location for business
• The UK's leading practical, community scale test centre for energy efficient living and low carbon transportation
• To redistribute the benefits of its labours to the Corby community

The installation of electric vehicle charge points are just one part Electric Corby's vision. The Electric Corby branded charge points are part of the Plugged in Midlands Scheme, which is now operated by Polar. A Polar Access Card or Smartphone is required to access the charge points.

Right: *A Renault Zoe*

10.7 Energise Network

Geographical Area: South East, including Surrey, Sussex and Kent

• Website:	www.EnergiseNetwork.co.uk
• Charger Types:	Type 2 Fast and Rapid AC/DC Chargers
• Operated by:	Charge Your Car (CYC) for Energise
• Access:	Energise Access Card, Charge Your Car card, Charge Your Car Smartphone App, or Pay As You Go via the telephone – 0191 260 5625 – 24 hours a day
• Access Cost:	Free or £10 (one off fee, including VAT)
• Electricity Charges:	Varies. See actual charge point
• Telephone:	0191 265 0500 - office hours 0800 to 1800. Outside office hours for emergency assistance only
• Email:	Via www.energisenetwork.co.uk/contact

10.7.1 Details

The Energise Network is a coordinated scheme between the councils of Kent, Sussex and Surrey so creating an integrated charge point network for the South East. It is managed by Charge Your Car. An Engergise card is available or if you are a member of CYC or another CYC operated scheme your existing access card will work on CYC branded Energise points.

As part of the coordination of charge points across the South East, charge points provided by and are on Ecotricity, POD Point, Polar and Zero Carbon World schemes have been configured to accept the Energise card in the Energise area. The charge points from the other scheme providers in the Energise area are also shown on the Energise charge point maps. This is a fantastic piece of joined up thinking.

As the Energise network covers a number of international gateways, rapid chargers are being installed at the Ports of Dover and Newhaven as well as the Eurotunnel channel tunnel terminal at Folkestone. Energise chargers can also be found at London Gatwick Airport as well as key Southern Rail railway stations and along major roads, trunk routes and motorways in the region.

10.8 ESB ecars

Geographical Area: Republic of Ireland and Roaming to Northern Ireland

- Website: www.esb.ie/electric-cars/index.jsp
- Charger Type: Type 2 Fast and Rapid AC/DC Chargers
- Operated by: ESB
- Access: ESB ecars charge point Access Card or Smartphone App
- Access Cost: Currently free
- Electricity Charges: Currently free

- Telephone:
  From Ireland: 01 258 3799 or 1890 372 387
  From UK: 00 353 1 258 3799 - all 24 hours a day
- Email: ecars@esb.ie

10.8.1 Details

ESB is the state owned electricity company in the Republic of Ireland. Using €4.2 million (approximately £3.1million) of funding from the EU Trans–European Transport Network (TEN–T) Programme, ESB ecars have installed 41 rapid chargers across the Republic.

The aim of the scheme is to have a rapid charger at service stations and other prime locations every 60km (37 miles) on Ireland's main intercity inter–urban route road routes. This offers seamless re–charging and payment, regardless of location across the whole of Ireland (Republic of Ireland and Northern Ireland) so reducing barriers to movement, including trade and tourism. The TEN–T money is also providing seven fast public charge points at Irish Rail's Heuston Station, Kent Station Cork, Athlone, Newbridge, Kilkenny, Waterford and Dundalk Stations.

10.8.2 Roaming

An additional 5 rapid chargers have been installed across the border in Northern Ireland, in association with the Department for Regional Development Northern Ireland. You can also use the ESB ecars charge point access card or Smartphone App on ecarNI charge points in Northern Ireland.

If you are a UK resident and want to obtain an ESB ecars access card, then they will accept applications from outside of Ireland.

Above: *An ESB ecar Siemens QC45 tri-rapid charge point at Topaz Motorway Services near Cashel on the M8, junction 8. Picture by courtesy of Mark Long.*

10.9 GMEV – Greater Manchester Electric Vehicle

Geographical Area: Greater Manchester only

- Website: www.ev.tfgm.com
- Charger Types: Type 2 Fast and Rapid AC/DC Chargers
- Operated by: Charge Your Car (CYC) for GMEV
- Access: GMEV Access Card, Charge Your Car card, Charge Your Car Smartphone App, or Pay As You Go via the telephone – 0191 260 5625 – 24 hours a day
- Access Cost: Free or £10 (one off fee, including VAT)
- Electricity Charges: Currently Free

- Telephone: 0191 265 0500 - office hours 0800 to 1800. Outside office hours for emergency assistance only
- Email: pluggedin@tfgm.com

10.9.1 Details

Launched in July 2013, Greater Manchester Electric Vehicle or GMEV is run for Transport for Greater Manchester (TfGM) by Charge Your Car (CYC). The aim of TfGM, is to promote sustainable transport across the Greater Manchester region. This includes electric vehicles as well as other forms of transport including trams, buses, trains and cycling.

TfGM have always been a forward thinking organisation. Previously known as Greater Manchester Passenger Transport Executive (GMPTE), they were responsible for the introduction of clean electric trams to the streets of Manchester in the form of the Metrolink system way back in 1992. Metrolink was the first new tram, or light rail scheme in the UK for decades. TfGM also operate a number of electric vehicles in its own fleet including a Nissan Leaf, Toyota Prius hybrid and a Renault Kangoo Z.E. electric van. In addition when it comes to buses, Greater Manchester was one one of the first areas to have diesel/electric hybrid buses on the streets and more recently has introduced pure electric Optare "Metroshuttle" buses that operate around the city centre.

The GMEV electric vehicle charging points are located at various car parks across the region, including Metrolink tram Park and Ride sites as well as on street locations and private (NCP for instance) car parks. The GMEV access card, which costs a one off £10, will allow charging access on any Charge Your Car (CYC) charging scheme across the UK. Alternatively you can use the pay by phone option or Smartphone App to access a charge point.

Above: A Nissan Leaf charges at a Type 2 charge point at the Trafford Council offices.
Left: A Mitsubishi Outlander PHEV charges at the GMEV Type 2 charge point in the car park under the arches of Manchester Central Exhibition Centre.

10.10 Greener Scotland / ChargePlace Scotland

Geographical Area: Scotland only

• Website:	www.GreenerScotland.org
• Charger Type:	Type 2 Fast and Rapid AC/DC Chargers
• Operated by:	See Charge Your Car and Ecotricity
• Access:	See Charge Your Car and Ecotricity
• Access Cost:	See Charge Your Car and Ecotricity
• Electricity Charges:	Varies
• Telephone:	See Charge Your Car and Ecotricity or 0808 808 2242
• Email:	See Charge Your Car and Ecotricity or Via website

10.10.1 Details

Greener Scotland seems more to be a promotional website for electric vehicles and charge points in Scotland rather than an actual "bona fide" charge point scheme as such. From investigation, both Charge Your Car and Ecotricity seem to feature in the charge point mapping provided on the Greener Scotland website, but no details are offered on how you obtain a membership card for either scheme. Charge Your Car is offered as a helpline telephone number in case of difficulty, which is not much good if you are at an Ecotricity charge point! So if you are travelling to Scotland with an EV, it would appear you will need both a CYC card and an Ecotricity card for convenience. However, if you look at other charging maps, you will also see that there are Polar and ZeroNet charge points in Scotland as well. So as ever, you need a wallet full of access cards.

Greener Scotland is sometimes also referred to as ChargePlace Scotland. From investigation of locations listed as ChargePlace Scotland, they all appear to be CYC operated charge points. Confusing? Most definitely.

10.11 Hackney

Geographical Area: Hackney, London

• Website:	www.ChargePointGenie.com
• Charger Types:	Rapid AC/DC Chargers
• Operated by:	ChargePoint Genie
• Access:	Genie Smart Card, or Genie Smartphone App
• Access Cost:	£20 including VAT per Annum.
• Electricity Charges:	See below
• Telephone:	020 3598 4087 – 24 hours a day
• Email:	GenieSupport@ChargePointServices.com

10.11.1 Details

ChargePoint Genie provides the smart card access for three rapid chargers in Hackney, London. These are located at:

- Bentley Road, Dalston
- Calvert Avenue, Shoreditch
- Reading Lane, Hackney

When the rest of London is predominantly on the Source London scheme, it is somewhat odd that Hackney council chose ChargePoint Genie access cards for these three rapid charge points. Additionally, unlike other charge providers that allow access across a variety of schemes with the one card, a Hackney Genie Smart Card won't allow you access to the SSE chargers also operated by ChargePoint Genie in Hampshire, Cornwall, or at Eynsham, in Oxfordshire. This seems utterly ridiculous and is at odds with what the other charge point scheme providers allow.

Left: *A CYC branded APT Technologies tri-charger rapid charge point at Cathedral car park in Inverness. The hatched markings are a bit confusing and the charge cable would not reach the EV designated space on the left! From left to right, the connectors are CCS, Type 2 fast and CHAdeMO.*

10.12 inmotion!

Geographical Area: South Yorkshire – Barnsley, Doncaster, Rotherham and Sheffield

- Website: www.evinmotion.co.uk
- Charger Type: Type 2 Fast and Rapid AC/DC Chargers
- Operated by: npower Electric Vehicles
- Access: No card required
- Access Cost: Currently Free
- Electricity Charges: Currently Free

- Telephone: 0800 294 3568 – 0900 to 1700 Monday to Friday
- Email: None given

10.12.1 Details

inmotion! is a partnership between Barnsley, Doncaster, Rotherham and Sheffield councils as well as South Yorkshire Passenger Transport Executive. It has currently installed 13 charge points at locations across South Yorkshire, with npower Electric Vehicles administering the scheme.

The free to use charge points are a mixture of fast and rapids and no card is required to access them! The signage on the charge point I used did make reference to an access card though. However, I ignored this, connected the cable to my car and then to the charge point and it all worked without issue. As ever, some locations do charge for the actual parking, separately to the charge point use. So check before you walk away!

10.13 Just Park

Geographical Area: Nationwide

- Website: www.JustPark.com
- Charger Type: Various, depending on provider
- Access: Varies, depending on provider
- Access Cost: Varies, depending on provider
- Electricity Charges: Varies, depending on provider
- Telephone: Various. See below

- Email: Various. See below
- Operated by: Various. See below

10.13.1 Details

Unlike the majority of other schemes presented here, Just Park is not a provider or manager of charge point schemes as such. What it is though is a very simple and clever concept utilising the power of the Internet and modern mobile telecommunication in the form of the Smartphone, to empower individuals.

Tenants with car parking space on driveways or in garages etc can rent this space out to drivers who are looking to park their vehicle and want a recharge. If the person has a charge point fitted as well, then they can additionally rent that out, as part of the parking space charge or in addition to it. It is up to the location owner what they charge, in exactly the same manner as the larger schemes.

Like all these schemes, you need to pre–register your details and payment card with the website if you want to use a parking space. All parking payments are undertaken via the website, so you don't need to find money to pay to park. You can also pre–book your space, so taking the stress out of parking. You can easily search for a space by location and date, and then filter for electric vehicle charge point (referred to as Electric charging). Once you do find a charge point make sure it is compatible with your car type, noting that some people list a standard 3 pin plus socket under this option as opposed to a dedicated EV charging point. You can contact the space owner via the website to ask a question if need be. As already mentioned, there may be an additional charge levied by the space owner for recharging your car on top of the price for parking. Again, make sure you fully understand this, check if necessary. This is the only scheme that allows you to book a charge point online or via a Smartphone before arriving.

If you are a tenant or property owners with a parking space and charge point you just need to register all your details on the website. Make sure you clearly specify your charge point type and vehicles it is compatible with. If you don't know, then say this or go and find out, but please don't guess! If you are charging £X for the parking and £Y for the charge point electricity make this clear as well. You may want to charge a fixed amount for a recharge or an amount per car type. A Vauxhall Ampera or Chevrolet Volt (EREVs) will likely require less of a charge than (say) a Nissan Leaf or Renault Zoe (EVs).

Above: A Just Park tethered charge point, operated by a key switch. This is a Polar manufactured unit installed by British Gas.
Page 60: Clear road markings and signage at Meadowhall shopping centre, Sheffield.

10.14 Milton Keynes Crosslink

Geographical Area: Milton Keynes and surrounding area

- Websites: www.ChargeMasterPLC.com
                       www.PolarInstant.com
- Charger Type: Rapid AC/DC Chargers
- Operated by: Polar
- Access: Polar Access Card or Polar Instant Smartphone App
- Access Costs: Free, £12/month or £20/year
- Electricity Charges: Varies depending on the access scheme. See below

- Telephone: 0330 016 5126 – 24 hours a day
                       01582 400331 – 0830 to 1730 Monday to Friday
- Email: Via Website www.ChargeMasterPLC.com

10.14.1 Details

Milton Keynes Crosslink is a Milton Keynes Council initiative to make the city a low–carbon and emission–free centre and to encourage easier access to the Milton Keynes regional shopping centre and recreational facilities. The scheme, which is funded by OLEV and Chargemaster (no MK council money was spent), is a network of 14 Rapid AC/DC electric vehicle charging points located at strategic locations across the region. The intention is to create "electric roads" so enabling pollution free mobility between Milton Keynes and its surrounding area including Bedford, Buckingham, Cambridge, Cheltenham racecourse, Silverstone Circuit and Oxford.

Above: A rather anonymous DBT-CEV tri-charger rapid charge point sits in the corner of the car park at Cheltenham racecourse. There is no signage to find it either!

10.15   Plugged–in–Midlands

Geographical Area: East and West Midlands
Worcester to Lincoln and from Stoke on Trent to Kettering

- Website: www.PluggedInMidlands.co.uk
- Charger Type: Type 2 Fast and Rapid AC/DC Chargers
- Operated by: Polar
- Access: Polar Access Card or Polar Instant Smartphone App
- Access Costs: Free, £12/month or £20/year
- Electricity Charges: Varies. See below

- Telephone: 0845 838 0551 – 24 hours a day
                           01582 400 331 – Office Hours 0830 to 1730
- Email: PiM@ChargeMasterPLC.com

10.15.1   Details

The Plugged–in–Midlands scheme was originally operated by POD Point and installed over 700 electric vehicle charging points across the East and West Midlands. It was one of the original 8 'Plugged–in–Places' projects, which was financially supported by OLEV.

In July 2015, the 870 charge points on the Plugged–in–Midlands scheme changed from POD Point operation to Polar operation, although for reasons not defined, some charge points were migrated over to Charge Your Car at this time, just to add further confusion and frustration to EV drivers.

Above: A Nissan e–NV200 Van

10.16 POD Point

Geographical Area: Various areas, nationwide

- Website: www.POD-Point.com
- Charger Type: Type 2 Fast chargers only
- Operated by: POD Point
- Access: RFID Access Card (now discontinued) or POD Point Open Charge Smartphone App
- Access Cost: Varies. See individual schemes
- Electricity Charges: Varies. See individual schemes

- Telephone: 0333 006 3503 or 0207 247 4114
                          0800 to 2000 – 7 days a week
- Email: enquiries@POD-Point.com

10.16.1 Details
POD Point design and manufacture a range of hardware products, as well as running the POD Point Network. They also supply the management software so allowing charge point providers to manage their network of charge points. In addition to this, POD Point offers a turnkey solution to businesses to survey, design, install and even manage an electric vehicle charge point solution, including maintenance. This means the benefits of electric vehicle infrastructure can be added to a business with the minimal of hassle, so allowing the business a number of green benefits. These include:

- Implementing their own electric vehicle fleet
- Meet environmental commitments
- Attract "green" EV driving customers
- Increase customer dwell time and average spend

10.16.2 POD Point Open Charge
In September 2015, POD Point launched "Open Charge" in response to the issue of membership cards and charge point accessibility. Open Charge is a new type of EV charge point and does away with the need for membership or an RFID smart access cards but you do need the POD Point Open Charge Smartphone App. You cannot use the RFID card on these charge points. There are no monthly fees with Open Charge. To use Open Charge you:

- Park at a POD Point charge point that works on Open Charge. Not all POD

Point charger points are configured to Open Charge yet, as this is an ongoing conversion process.
- Plug in your EV. The charge starts immediately
- You then have 15 minutes to access the POD Point Open Charge web or Smartphone App to confirm the charge you have initiated. The 15 minute time window allows you to:

  1. Get a Smartphone signal if you are in a poor coverage area, including in an underground car park
  2. Get to a computer if you are heading to the office, coffee shop etc
  3. Get out of the rain/snow etc

POD Point has also worked on improving the reliability of the Open Charge EV charge points. Chargers start immediately on connection to an EV and if they fail to communicate they will continue to charge the EV. This is a fail safe in the favour of the EV drivers who needs a charge. The issue of the drivers charge cable being trapped in the charge point has also been addressed as it is released from an Open Charge point when the car stops charging.

Above: *A POD Point charge point at Haddenham & Thame Parkway railway station.*

10.17 Polar

Geographical Area: Various areas, nationwide

- Websites: www.ChargeMasterPLC.com
                        www.PolarInstant.com
- Charger Type: Type 2 Fast and Rapid AC/DC Chargers
- Operated by: Polar
- Access: Polar Access Card or Polar Instant Smartphone App
- Access Costs: Free, £12/month or £20/year
- Electricity Charges: Varies depending on the access scheme. See below

- Telephone: 0330 016 5126 – 24 hours a day
                        01582 400331 – 0830 to 1730 Monday to Friday
- Email: Via Website www.ChargeMasterPLC.com

10.17.1 Details

Polar is a brand name of Chargemaster Plc, who both manufacture, install and manage electric vehicle charging points across the UK. Polar offer three options for charge point access, depending on your likely usage. These are:

- Economy Plus Tariff: Monthly subscription of £12 with an amount of inclusive minutes, then you pay for what you use on top of that on a pro–rata basis
- Standard Tariff: Annual subscription of £20. You then pay for what you use, with a minimum charge of one hour
- Instant Smartphone App. Pay As You Go, with a minimum charge of one hour

Charges are cheapest for the Economy Plus Tariff and the most expensive for the Instant Smartphone access. Rates inclusive of VAT are currently:

- 13 amp Socket per hour   £ 0.95 / £1.00 / £1.20
- Type 2 Charger per hour   £ 1.45 / £1.50 / £1.70
- Triplex Rapid Charger per 30 minutes £ 6.00 / £ 7.00 / £7.50

Chargemaster, through their Chargelease arm, also offer businesses the opportunity to lease a 7kW, key operated, fast charger unit for as little as £45 month. This includes the installation and ongoing maintenance costs, making it really easy for a business to "go green".

10.18 Recharge

Geographical Area: Liverpool City Region, Merseyside, Cheshire West and Chester

- Website: www.MerseyTravel.gov.uk/Recharge
- Charger Types: Type 2 Fast Only
- Operated by: Charge Your Car (CYC)
- Access: Charge Your Car Access Card, Charge Your Car Smartphone App, Pay As You Go via the telephone 0191 260 5625 – 24 hours a day
- Access Cost: £20 including VAT per Annum. See below
- Electricity Charges: See individual schemes

- Telephone: 0191 265 0500 - office hours 0800 to 1800. Outside office hours for emergency assistance only
- Email: admin@ChargeYourCar.org.uk

10.18.1 Details

Led by Merseytravel, the Merseyside Passenger Transport Executive, and part funded by OLEV, Recharge has rolled out 46 charge points across the Merseyside and West Cheshire regions. The charge points are currently free to use, apart from any actual car parking charges, and are operated by CYC. This is very fortunate, as the adjoining GMEV scheme is also operated by CYC, so providing seamless charging opportunity across the Greater Manchester and Liverpool areas. Joined up thinking for once!

The Recharge partners, who have made the scheme happen, include:

- Cheshire West and Chester Council
- Halton Borough Council
- Knowsley Metropolitan Borough Council
- Liverpool City Council
- Sefton Metropolitan Borough Council
- St Helens Council
- Wirral Metropolitan Borough Council
- Merseyrail, Northern Rail and Virgin Trains

Right: *An APT Technologies manufactured "evolt" Standard Design Post Type 2 fast charger on the Recharge (CYC) scheme at Earlestown in Merseyside.*

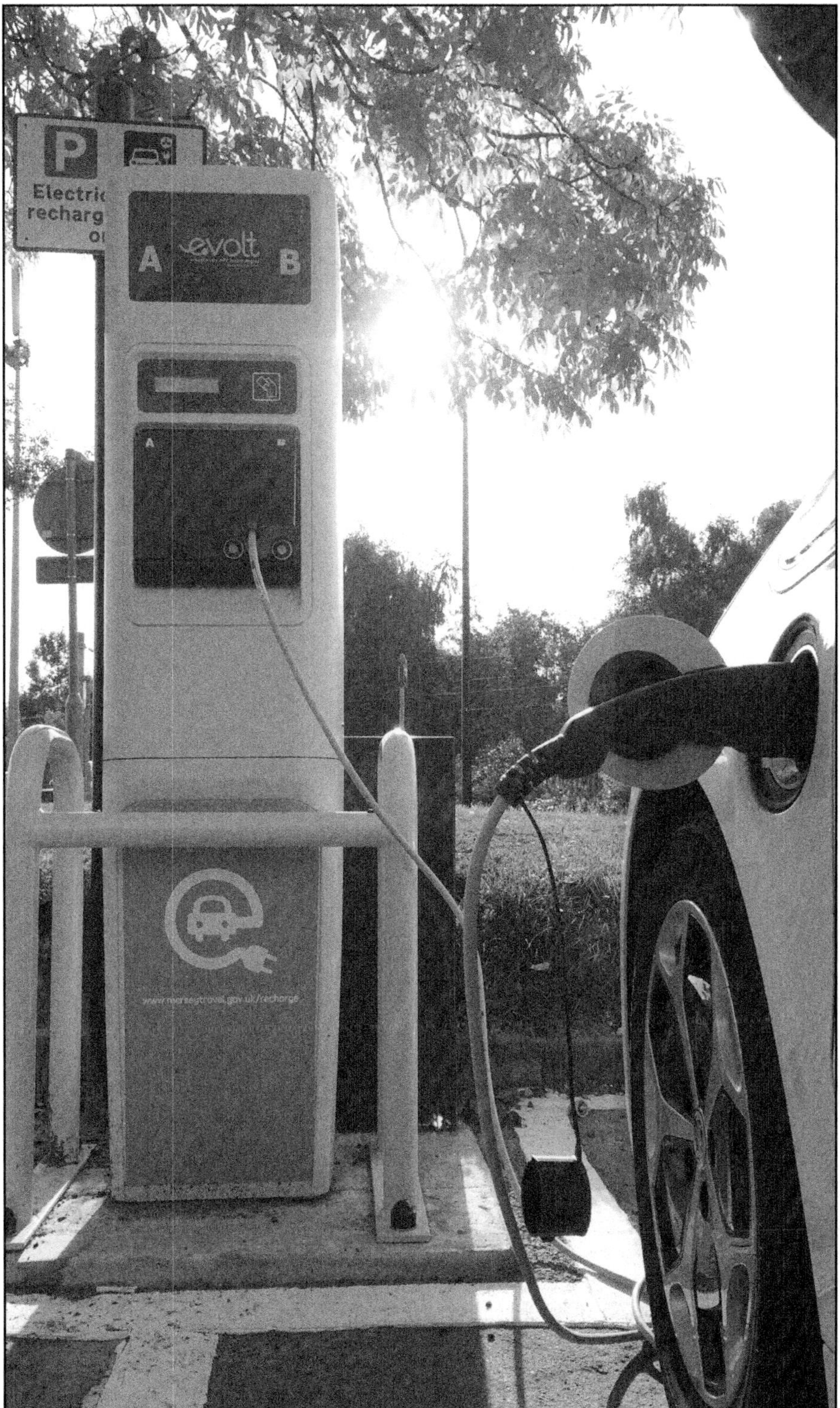
P
A
evolt
B
www.merseytravel.gov.uk/recharge

10.19 Source East

Geographical Area: East of England (East Anglia)
Roaming Agreement: Plugged in Midlands and Source London

- Website: www.SourceEast.net
- Charger Type: Type 2 Fast and Rapid AC/DC Chargers
- Operated by: POD Point
- Access: Source East membership card
- Access Cost: £10/year (including VAT)
- Electricity Charges: Currently Free

- Telephone: 0845 519 8676 or 0207 247 4114
                        0800 to 2000 – 7 days a week
- Email: SourceEast@SourceEast.net

10.19.1 Details

Source East is the organisation responsible for electric vehicle charging points in East Anglia, providing a network of over 800 charging points. It aims to offer a charge point within 25 miles of all local residents and bushiness within the East of England.

Charging points are conveniently located at various publicly accessible places all over the East of England, including:

- Supermarkets
- Railway station
- Public car parks
- Outside local businesses
- Residential streets

The charging points have been installed at what Source East refer to as "hot–spots" within their geographical operating area, including in and around the following key locations:

- Bedford
- Cambridge
- Hertfordshire
- Ipswich
- Luton

- Norwich
- Peterborough
- Stansted Airport
- Thames Gateway South Essex

10.19.2 Roaming
Source East customers can also use the following networks

- Plugged–in–Midlands
- Source London

However, it should be noted that if you roam on another networks EV charging points, you will have accepted the terms and conditions of that network by default. You can also use some Ecotricity charge points that are in the Source East area and are labelled as such.

Right: *The label on the side of an Ecotricity charge point that is situated in the Source East area. A Source East, Source London, Plugged-In-Midlands or CYC card holder can also access and use this charge point. The cross scheme roaming access is not reciprocal and only works at labelled charge points, not all charge points.*

10.20 Source London

Geographical Area: London

- Website: www.SourceLondon.net
- Charger Type: Type 2 Fast and Rapid AC/DC Chargers
- Operated by: IER Bolloré Group
- Access: Source London Access Card
- Access Cost: £5 subscription fee (until 31/01/16)
- Electricity Charges: Currently Free

- Telephone: 0203 056 8989 – 24 hours a day
- Email: Via Website www.SourceLondon.net/contact

10.20.1 Details

Source London was launched by Transport for London (TfL) in May 2011 with the aim of providing an integrated charging network for electric vehicles. Initially TfL worked with the various London Boroughs and private partners to install charge points in the capital. In September 2014, the scheme was transferred to the Bolloré Group to run and they are working with the various London boroughs to either install new charge points or retrofit existing points

In addition to the London Boroughs, a number of private partners are also working with Source London to install charging points. These include hospitals, shopping centres and private car parks. All charging points installed privately will be branded with the Source London logo to show it is part of the scheme.

TfL continues to promote the uptake of electric vehicles through a range of initiatives including no Congestion Charge which could save a driver up to £2,990 for cars and vans that are driven into central London, and the introduction of lower emission buses and taxis.

Source London also allows you to reserve a charge point via its website or mobile phone app. Currently over 1,300 are installed and the intention is 4,500 more will be installed by 2018.

10.21 Source West

Bristol, Bath, North East Somerset, South Gloucestershire, North Somerset and Gloucestershire.

- Website: www.SourceWest.info
- Charger Type: Type 2 Fast and Rapid AC/DC Chargers
- Operated by: Charge Your Car
- Access: Charge Your Car Access Card, Charge Your Car Smartphone App, Pay As You Go via the telephone 0191 260 5625 – 24 hours a day
- Access Cost: £20/year (also see below and Charge Your Car page)
- Electricity Charges: Currently Free for standard charges. A fee applies for rapid charges

- Telephone: 0191 265 0500 - office hours 0800 to 1800. Outside office hours for emergency assistance only
- Email: info@SourceWest.info

10.21.1 Details

Source West is the charging scheme that covers Bristol, Bath and North East Somerset as well as Counties of South Gloucestershire, North Somerset and Gloucestershire. Funded by the Local Sustainable Transport Fund through Bristol City Council, the scheme is operated by Charge Your Car.

To encourage uptake to the scheme by locals who live in the Source West area, a number of interesting incentives have been offered. These include:

- Free access cards for the first year for the first 350 local members. However, you will be expected to complete 4 questionnaires over the course of the year
- A subsidy towards the cost and installation of intelligent charge points for both home and work. However, to qualify for this subsidy, you are asked that you change to a "green" electricity supplier at the location the charger is installed.

For those outside the Source West area, an access charge card will cost £20/year.

10.22　SSE

Geographical Area: Hampshire, Isle of White, Cornwall and Oxfordshire (1x)

- Website: www.ChargePointGenie.com
- Charger Types: Type 2 Fast and Rapid AC/DC Chargers
- Operated by: ChargePoint Genie
- Access: Genie Smart Card, or Genie Smartphone App
- Access Cost: £20 including VAT per Annum.
- Electricity Charges: See below

- Telephone: 020 3598 4087 – 24 hours a day
- Email: GenieSupport@ChargePointServices.com

10.22.1 Details

SSE (formerly known as Scottish and Southern Energy) have been responsible for the installation of a number of charge points across the south of England. This has been undertaken in conjunction with various local councils, who have obtained the necessary government funding and have facilitated the installations. Whilst SSE have provided the installation project management and charge point equipment, Charge Point Services have provided the back office software, through ChargePoint Genie.

Like most schemes, there is an RFID access card charged at £20 a year as well as a Smartphone App. Electricity is charged at £0.30 per kilowatt/hour (kWh) used. You have to keep a credit balance in your account in order to use the charge points and there is a minimum top up purchase of £1.00. So not strictly speaking Pay As You Go, unlike some other schemes.

ChargePoint Genie also has some other interesting differences including:

- Charge initiation fee. You will pay £0.60 on a fast charge and £1.80 on a rapid charge, just for commencing the charge.
- Overstay charge. You will pay £10.00 if your vehicle is left on charge for more than 4 hours (fast charge) or after an hour on rapid charge. For a rapid charge this is repeated every hour.
- ChargePoint Genie have restricted the rapid charger to only deliver an 80% charge and then cut out. If you want to get that 20% extra so you have a full battery, you need to start the charge again, which will mean another initiation feed of £1.80!

The overstay charge has been levied to make sure that there is a turnover of people using the charge point and it is not hogged, especially the rapid version. The idea being that a driver who needs a rapid charge will not have to wait more than an hour to speak to the driver of the vehicle at the charge point.

Above: The lone SSE branded APT-Technologies "evolt" rapid Compact tri-charge point located at a BP petrol station on the west bound A40 near Eynsham in Oxfordshire, close to the junction with the B4449. From left to right, the connectors are CCS, Type 2 and CHAdeMO.

10.23 Tesla Supercharger Network

Geographical Area: Nationwide for Tesla cars only

- Website: www.TeslaMotors.com/en_GB
- Charger Type: Rapid 120kW for Tesla Model S Cars Only
- Operated by: Tesla Motors
- Access Cost: Free – included in the price of the car
- Electricity Charges: Free – included in the price of the car

- Telephone: 0800 756 9960 – 24 hours a day
                         01628 450 600 (office hours)
- Email: Supercharger–EU@TeslaMotors.com

10.23.1 Details

Tesla has started to role out a network of 120kW rapid superchargers. These 21 chargers (at March 2015) are installed around the UK, but are currently London centric, and are only suitable for the company's Model S car. They are not compatible with any other car, including the Tesla Roadster model, which was an all electric car based on the Lotus Elise. So unless you have a Tesla Model S, then the network is not any use to you.

However if you do drive a Model S, then you have a purpose made charge point network just for you in the UK and into mainland Europe. In addition, you will have no worries about charge socket compatibility issues or access card acceptance. Better still the electricity is free, as it is factored into the cost of the car, which will conveniently find you a supercharger when you are out and about. Even better is that following a high profile legal spat between Ecotricity and Tesla, which was settled in July 2015, superchargers are now set to appear at UK motorway services.

10.24 Think Travel/Think Electric

Geographical Area: Gloucestershire. Part of the Source West scheme

- Website: www.ThinkTravel.info/car/electric–vehicles
  www.SourceWest.info
- Charger Type: Type 2 Fast and Rapid AC/DC Chargers
- Operated by: Charge Your Car
- Access: Charge Your Car Access Card, Charge Your Car Smartphone App, Pay As You Go via the telephone – 0191 260 5625 – 24 hours a day
- Access Cost: £20/year (also see below and Charge Your Car page)
- Electricity Charges: Currently Free for standard charges. A fee applies for rapid charges
- Telephone: 0191 265 0500 - office hours 0800 to 1800. Outside office hours for emergency assistance only
- Email: enquiries@thinktravel.info
  info@SourceWest.info
  admin@ChargeYourCar.org.uk

10.24.1 Details

Gloucestershire County Council has been responsible for installing an initial 12 electric vehicle charging points in and around Cheltenham and Gloucester. The charge points are part of the Source West Scheme, operated by Charge Your Car. A Charge Your Car access card will operate it.

Left: Two Tesla Superchargers stand next to a Chargemaster branded fast charger at the Park Royal Hotel, Stretton, which is just to the south of Warrington. Whilst the hotel is near the M56 (J10), it is located a good distance away, off the B5356 Stretton Road and the charge point is not signposted. Additionally, when you arrive at the hotel, the Superchargers are located round the back of the hotel, literally as far as you can drive in the car park, again without signage. The large white box and the smaller grey box behind the fast charger are part of the supercharger installation. Interestingly, the fast charger is switch operated and would appear not to need a card to operate it. It also does not appear on Chargemaster maps.

10.25 ZeroNet Network by Zero Carbon World

Geographical Area: Global

- Website: www.ZeroCarbonWorld.org
- Charger Type: Various, depending on provider
- Operated by: Various, depending on provider
- Access Cost: No card required.
- Electricity Charges: Free or Pay As You Go

- Telephone: 01225 667544 (office hours)
- Email: Via contact form –
                        www.ZeroCarbonWorld.org/contact–us

10.25.1 Details

Zero Carbon World Limited is the only charity company to feature among the providers of electric vehicle charging points. The aim of the organization is impressive, in that it wants to build an unrestricted national charging network, ZeroNet (www.ZeroCarbonWorld.org/zero–net) This is being achieved through actually donating charging stations to businesses as in the hotel and leisure industry. The only requirement is for the business to pay for the actual installation of the charge point and to operate it.

Unlike other electric vehicle networks, to use ZeroNet you do not need to register for a smart card or pay any annual fees. You only need to turn up to a charge point and plug in, paying any fees for parking or electricity as required at that location. Some sites may be totally free. All in all, a much simpler, much fairer and a totally inclusive system for anyone to use.

As mentioned, the actual charge points will either be Pay As You Go or free, depending on the business that has installed it, what it's business model is and what it wants to achieve. Free may encourage more customers to visit the business, who may stay longer whilst charging their car. Pay As You Go may be for "needs must" recharges and to "tick the proverbial green credential box" for businesses.

In addition to the ZeroNet charging points, Zero Carbon World have also developed Open Charge Map (OCM). This is a database containing information on electric vehicle charging points across the world. It lists useful charge point information including:

- Output/connections
- Operational Status
- How to find it

The information is maintained and updated by a community of volunteers who keep the information up to date. The aim is to create a single free global database of up to date information, covering all providers and so removing the need for multiple private databases/maps, which may only reference one provider of charge point.

Zero Carbon World is allowing developers to freely use the Open Charge Map database to create interactive maps, smart–phone/table Apps or even in Sat–Nav systems, for use by anyone who has an electric vehicle.

Above: *A Zero Carbon World charge point at the beautiful 18th Century Bloomfield House eco guest house in Bath. On the right a Type 2 and on the left a 3 pin socket, both operated by switches above the sockets. Picture taken with thanks to Bloomfield House (www.ecobloomfield.co.uk)*

11.0 Businesses with Charge Points

A number of businesses around the country have taken the initiative to have electric vehicle charge points installed in their car parks. Public facing businesses that I am aware of are listed below, in alphabetical business order, along with the infrastructure scheme provider. Not all business locations necessarily have charge points installed.

- Accor Hotels – Polar
- Asda – Polar
- Best Western Hotels – ZeroNet Network (Zero Carbon World)
- BMW Dealership – Charge Your Car
- BP Petrol Station – SSE/ ChargePoint Genie
- Chiltern Railways – POD Point
- Esso – SSE/ChargePoint Genie
- Extra Motorway Services – Ecotricity
- Ibis Hotels – Polar
- Ikea – Ecotricity
- Intu Trafford Centre – GMEV

- Little Chef – Polar
- Marriott Hotels – ZeroNet Network (Zero Carbon World)
- Meadowhall Shopping Centre – inmotion!
- Metrolink Park and Ride Sites (tram system, Greater Manchester) – Charge Your Car
- National Trust – ZeroNet Network (Zero Carbon World)
- Nissan – POD Point

- Novotel Hotels – Polar
- Peugeot Dealership – Charge Your Car
- Q Hotels – Polar
- Roadchef Motorway Services – Ecotricity
- Sainsbury's – POD Point and Polar
- Southern Railways – Charge Your Car
- Tebay Services – ZeroNet Network (Zero Carbon World)
- Tesco – POD Point
- Toyota Dealerships – Polar
- Welcome Break Motorway Services – Ecotricity
- Waitrose – Polar

It should be noted that some of these charge points are only available when the business is open or if you are staying at, or trading with the business. For example, you cannot use an Ecotricity charge point at an Ikea on Easter Sunday, as the gates to the car park are well and truly locked!

In addition, a number of car dealerships selling EVs or EREVs have charge points fitted to support the cars they are selling. However, I am not sure they would welcome you with open arms if you turned up in a make and model of EV they did not sell, no matter how desperate you were for a charge! You can check for any access restrictions at an EV charge point by using one of the various online mapping services.

Above: An Ecotricity DBT-CEV tri-charger rapid charge point and Type 2 charger at a Keele Welcome Break services on the southbound M6.

Left: GMEV Type 2 chargers at the Intu Trafford Centre in Manchester.

12.0 The Future?

We live in exciting times. Things are changing rapidly all around us, all the time. Electric vehicles are still really in their infancy, mainly due to the battery capacity. But battery capacity has come on in leaps and bounds in the last few years. The idealistic, wish-list perfect battery that everyone is chasing is:

- Small size – no bigger than a fuel tank or smaller
- Big capacity – range of an ICE car or greater
- Fast recharge time – comparable with refuelling an ICE or quicker
- No degradation over time due to rapid charging

Recent developments in lithium-air batteries look promising for EVs. Using a spongy graphene electrode and a new chemical reaction they can store much more energy than today's lithium-ion batteries. But the technology maturing for commercial use may still be some 10 years away.

It is not just the battery development that is important. Charging technologies are evolving as well. Smaller battery capacity but with better charging opportunities may be the next development instead, as discussed further below.

12.1   Induction Charging
Using induction charging technology, vehicles can now be charged using loops embedded in the road surface of a car park. When the vehicle is parked above an induction loop it will automatically be recharged, with no need for a physical connection. The need for different charge plugs/sockets and the incompatibility this causes will be eliminated. Large and heavy batteries will be removed as well, as it would be possible to charge every time you park you car.

Induction charging on the move is also to be trialled. The induction loops will be fitted in the road and by driving over them in your suitably fitted EV, the battery will be charged on the move.

12.2   Battery Swapping
As well as rapid charging technologies, some are favouring battery swapping technology. Just as you would swap out AA batteries in a portable radio, a battery swap under a vehicle is possible. The vehicle is taken to a special location where the nearly depleted battery is released from the bottom of the

car and a new battery is installed. Tesla have designed their cars to accept this and are currently experimenting with it. Israel is also keen on developing this technology further.

12.3 Solar

Solar is another technology that could easily be applied to a car and I am surprised there are not more around at the moment. As solar panels become more efficient, cheaper, lighter and even flexible, a thin film solar panel could easily be mounted on the roof of an EV, in the same manner as a panoramic glass roof. This would then harness any available sunlight it gets to help supplement the battery charge, in addition to charging by more conventional means. It is unlikely that a solar panel on the roof will harvest enough power from the sun in the UK to fully recharge an EV, but every little helps the traction battery.

Interestingly, Ford in the United States unveiled a prototype C-Max Solar Energi Concept car back in early 2014. The design was reported to track the sun and could harvest solar power during the day equal to a four hour battery charge, which would give around 21 miles of range. Not so sure the same range would be possible in the cloudier UK however!

12.4 Other Low Carbon Emission Technologies

In addition to battery/motor and perhaps an ICE for an EV or EREV, there are other forms of low carbon emission vehicle technologies in the offering at the moment. Some are listed below and these could be what the future holds instead of an EV as we know it.

12.4.1 Hydrogen Fuel Cell

More interestingly though is the steps the likes of Toyota have taken with hydrogen fuel cell technology. Running on hydrogen, that can be converted to make electricity with a waste by product of water, hydrogen vehicles can be refuelled like a normal petrol vehicle, in a similar time. It also offers a long range for the amount of hydrogen put in, again comparable with an ICE.

The problem with hydrogen though is it is a very volatile substance. The R101 airship was filled with hydrogen and look what happened to that! With today's modern materials, the fuel cell is designed to be inherently safer than a cloth airship filled with hydrogen gas, as the tank is wrapped in carbon fibre and is safely buried away in the vehicle, so that issue should have been addressed.

Below: *A Toyota Mirai hydrogen fuel cell car at the Low Carbon Vehicles 2015 event at Millbrook Proving Ground in Bedfordshire.*

The other issue with hydrogen is making and supplying it. This usually requires an industrial process, which is very energy hungry and the necessary distribution infrastructure to be created, which really is no different from refining fuel and getting it to the filling stations. Fortunately though, micro hydrogen stations have been developed and put into production, using a wind turbine or other renewable energy source to create the hydrogen at the point of refilling. Sheffield based ITM Power have installed such a hydrogen refuelling station at the Advanced Manufacturing Park near the M1 in Rotherham. In a different approach to the problem, scientists at London's Imperial College Energy Futures Lab have recently been successful in developing a way to continuously produce hydrogen from algae!

Unlike a conventional electricity generating power station, you cannot switch renewable energy on or off. If it is windy or sunny, you are generating electricity, whether if you want to or not. If it is not windy or sunny then you get nothing. Traditional power generation then has to fill in the demand gap. You also cannot store large quantities of electricity easily. But hydrogen production may be one missing piece of the puzzle for renewable energy that will also benefit transportation, by making and storing hydrogen, that can be later converted back to electricity via the hydrogen fuel cell.

12.4.2 LPG/Petrol

Liquid Propane Gas (LPG) conversions have been around for many years. An LPG tank is fitted to the vehicle and a mix of LPG and petrol is burnt in the ICE, to make the vehicle go. The amount of LPG is varied, with usually more petrol being used on start up. LPG was always marketed as a way to reduce your vehicles running costs, with a litre of LPG costing around half the price of unleaded. Additionally though, LPG can reduce your vehicle emissions, so is "greener" than a 100% petrol vehicle. Surprisingly a conversion to an existing car costs less than £2,000 and in mainland Europe some of the car manufacturers offer an LPG/petrol model as part of the range. A reputable converter will provide the necessary warranties as well for the conversion which will not impact your existing warranty on the vehicle. An LPG conversion drives and handles exactly as per a standard ICE and the blending of LPG and petrol in the ICE is all done automatically, without any manual intervention required.

I was fortunate enough to get the opportunity to drive an Autogas Ford Focus LPG car around the Millbrook proving ground in Bedfordshire. It drove and handled as a Ford Focus should do, and I have driven a lot of them over the years. If you run out of LPG because you cannot get to a suitable LPG filling station, then the petrol keeps you going. So no range anxiety. A number of black cab taxi conversions are currently being trialled using older models, with a new petrol engine and an Autogas LPG conversion, in an attempt to reduce carbon emissions in city centres, due to dirty diesel cabs.

Above: An Autogas LPG converted TX4 "Black Cab" taxi and Ford Focus at the Low Carbon Vehicles 2015 event at Millbrook Proving Ground in Bedfordshire.

12.4.3 Hydrogen/Diesel

Liverpool based Ulemco have developed a hydrogen/diesel hybrid system. This works on a similar idea to an LPG conversion with a petrol engine. The technology allows lower vehicle emissions by mixing hydrogen in with the diesel fuel, so lowering the amount of diesel used at certain speeds and so cutting exhaust emissions. One or two hydrogen fuel tanks are retrofitted fitted into the vehicle, so this is targeting reduced emissions in the commercial van or lorry market. With careful design and fitting, no additional "business" space is taken up, as there is no reduction in load capacity, for instance by installing the required equipment in the back of a panel van. A refuse truck is also being converted as a test case.

Gas and fuel (petrol or diesel) mixes are all well and good for short term emission reduction but one of the main aims must be to remove the need for any type of hydrocarbon fuel requirement.

12.4.4 Bio-methane

Bus operator First Bristol, Bath and the West, along with Geneco, trialled a bio-methane powered bus between November 2014 and November 2015 in the Bristol area. The bio-bus uses a mixture of methane and propane, the former created from sewage and food waste. The methane bio-gas is created through anaerobic digestion. This is a process where oxygen starved bacteria breaks

Above: Müller Wiseman have been trialling dual fuel lorries. These use Liquefied Natural Gas (LNG) which is predominantly methane. The Volvo Fm13 is seen at the Low Carbon Vehicles 2015 event at Millbrook Proving Ground in Bedfordshire.

down the biodegradable waste to produce a methane rich bio-gas. The biogas then has the carbon dioxide and other impurities removed before propane is added. The resultant gas, manufactured by Wessex Water subsidiary company Geneco, is compressed into a liquid in a similar manner to that of LPG and hydrogen previously mentioned. The bio-gas is stored in tanks on the bus roof, so maximising space in the bus. The 40 seat bus is reported to emit up to 30% less carbon dioxide (CO_2) than similar diesel buses.

The successful trial has resulted in plans to be drawing up jointly by First and Wessex Bus to apply to OLEV for a £2.5 million grant to support the ordering of 130 bio-gas buses. This is for 110 double deckers for First and the rest for Wessex Bus.

12.5 Hindsight: A Wonderful Thing

With hindsight, the electric vehicle charge point role out in the UK has not really been the best from the end users point of view. Vested interests, rushed installations, the need to spend budgets quickly due to false spending deadlines, poor location decisions, "pounds in the grounds" PR sound bites for politicians and councillors, incompatible infrastructure with earlier EV models, fragmented scheme providers, lack of scheme roaming agreements, sky high prices for recharging (in some instances but not all) and legal disputes have not helped further the cause of the electric vehicle. This needs to change if we, as both the UK and the world, are to succeed in reducing our dependency on oil. The Republic of Ireland, being a smaller market and using EU money, seems to have pulled it off properly, proving it can be done where there is a will to deliver.

12.6 So what is the ultimate EV Charge Point?

As this book is primarily about EV charging points, the proverbial $64,000 question has to be asked of what is the ultimate EV charge point? Well these are my thoughts on it, including what changes need need to be made to the actual EVs:

12.6.1 Standardisation of vehicle charge point socket

Whilst we could have vehicles offering a number of standard socket options and charge points offering a number of different connectors options, a common standard for the interface charge point connection is required. Just as per a petrol pump and car filler. I acknowledge that we do of course have a petrol and diesel pump versions, all for valid reasons, along with larger truck nozzle version. Tesla seem to have close to this in Europe with a modified

Type 2 charge connector that will fast and DC rapid charge. But this differs from what is on their US Model S, meaning left hand drive models in the US are physically different from European modes!

12.6.2 Longer Battery Range, Faster Battery Recharge Times and Lower Battery Cost

Longer battery range, faster recharge times and lower cost is of course the holy grail for an EV and we are getting closer to it all the time. Tesla have of course achieved two out of three so far with the Model S, but it is a large vehicle at a high price. Battery range needs to be at least 200 miles to be more appealing to the masses, to take into account heating/cooling needs and real life driving, so providing a decent driving range of at least 160 miles. This would be to give 2 hours of motorway driving at 70mph plus some back up range, before a fast charge is required. Of course battery ranges on a par with ICEs (300+ mile range) would be great, but with a better spread of rapid charge points it is not really necessary.

12.6.3 Variable Recharge Times for EVs

Electric vehicles should have the ability to be recharged at different rates so requiring lower power requirements. My Ampera will recharge from empty in 8 hours from a 3 pin plug or 4 hours from a fast charger. But I don't always need it to recharge in 4 hours. On a fast charge, it draws 3kWh. If it could recharge in 8 hours, and draw say 1.5kWh, I could probably recharge it fully from my 3kW solar system, over a day. Also by taking longer to recharge, less demand is put on the electricity grid. Longer, slower recharges are also better for your EV battery life.

12.6.4 The right sort of charger at the right place

A rapid charger should be placed at transient locations, like petrol stations and motorway service stations. They should be able to recharge a car to full in 30 to 45 minutes (or of course less!), the time I usually estimate to stop, use the facilities and have something to eat. Rapid chargers in town/city centre pubic car parks are not such a good idea, if typically the person who will be parking at such a location may be parking for an hour plus. At these locations we need to have semi–rapid chargers, that will recharge the vehicle in say 4 hours, or whatever the average parking time is calculated at. Or if you know you will spending 2 hours parked, then you can set the recharge rate to recharge your car to full in that time. At commuter car parks for railways stations or park and ride bus/tram locations, semi-fast chargers are required. These will be set to recharge the vehicle in 8 hours, or the duration that someone will be typically parked up at the site. The one size fits all charge point approach is not the right answer. Modern power electronics and software will provide the answer, if joined up thinking between EV and charge point manufacturers is had.

12.6.5 24/7 Charger Access

All chargers should be accessible 24/7/365. There should be no restrictions to accessing them if they are public facing, especially if they are rapid chargers.

12.6.6 Standardised Electricity Charges

Some chargers offer free electricity, which of course is fantastic for EV drivers. But this won't and cannot continue indefinitely. But if you have to pay, the price must be:

- A fair electricity price and of course cheaper than an equivalent petrol mile
- A standardised price across the UK for the given charge rate
- Clearly displayed price – just like at petrol stations
- Proportionate for what you use – no membership fee, connection charge, minimum charge or minimum charge time requirement. You should also be able to charge your EV to full if that is what you want, not just 80% capacity

Like petrol stations, EV charge points will be provided by private enterprises that want to make money. There is nothing wrong with that, as that is the world we all live in, but a fair price is necessary and unlike petrol prices, it should be consistent across the UK. Additional money making opportunities lie with the selling of refreshments etc whilst you wait for your EV to charge, at the charging stations or motorway services. Sighting charge points in high traffic areas and looking at ways to keep the costs down will help maximise profits.

12.6.7 A variety of Charge Point Providers

Like petrol stations, it is accepted there will be a number of different providers across the country. Unless the charge points are nationalised, and this is highly unlikely to ever happen, the different providers will probably all be using slightly different charge point equipment, but the basic operation and functionality will be the same throughout. However unlike petrol stations, there will be no competition between different charge points in the same area, based on price. The price will be standardised across the UK. Of course, this may need government intervention to enable it to happen, but the green benefits are repeated by all.

12.6.8 Standardised Payment Format

The need to join charging schemes, use Smartphone Apps or phone numbers to get a your EV charged is utter nonsense. You should be able to pay via a credit or debit card. As a charge will likely cost less than £10, this should be either a contactless payment or an insert and retrieve card, i.e., no PIN number is required, so no PIN pad. This will be in a similar way as with most road toll booths. If you want a receipt, or to see how many charges you have made, you should be able to log in to the website of the charge point provider that you used and register the payment card you used. Registering your card online will only link your EV charge point use (purchase) to you and not be a means to collect money. You will have to register at all charge points you use if you want to track what you have spent. Registering will also allow marketing opportunities and up selling by the charge point provider.

12.6.9 Charge Point Connections

As we have different EV vehicle types on the road, charge point locations need to be backwards compatible and all inclusive. This means every charger location should be able to recharge both Type 1 and Type 2 socket fitted vehicles. This can be achieved through the inclusion of a Type 2 socket on the charge point, so allowing a cable to be connected, or through the addition of a separate Type 2 fast charger. Whilst the recharging of EREV and plug-in hybrids at transient locations may seem counter productive, every mile of electric driving means a mile less of driving using fossil fuels.

12.6.10 Petrol Stations to become Energy Stations

With every EV bought, a petrol station's sales potential diminishes a little. Therefore they need to adapt to survive into the future. Petrol stations of course offer so much more than just petrol or diesel to us. They offer tyre inflation, car washes, confectionery, crisps, sandwiches, pies, newspapers, magazines, milk, food, cigarettes, drinks, coffee, cafe, toilets, lottery tickets, flowers, BBQ charcoal etc. Most of us, including non–car drivers, would really miss our local petrol station if it closed down. So petrol stations need to get with the times and embrace EV use by getting a rapid charger installed. I have only seen one such location in the UK, a BP station in Oxfordshire with a rapid charger. But the petrol station needs to have a cafe if it wants to maximise sales opportunity from a waiting EV driver. I would be happy to pay for a charge, buy a coffee, perhaps a snack and a newspaper or magazine whilst I sit and wait for my EV to charge up. On top of the charge, which the petrol station would be making some money from, there could be another £5 of sales potential as a minimum for a coffee and snack. If a family of 4 visited, perhaps

Below: An APT-Technologies tri-charge point located at a BP petrol station on the east/west A40 near Eynsham in Oxfordshire. The location is open 24 hours has a well stocked shop, cash machine, toilets and sells a variety of refreshments.

another £20 of sales potential and that is before I have seen that must have gadget reduced from £9.99 to only £3.99! Energy stations could also use their canopy roofs to fit solar panel so reducing running costs and helping to increase profits.

12.6.11 More Car Park Charge Points

As EV uptake expands, charge points should not be a specialist thing requiring online maps to find them. They should be a standard feature of every newly built car and retrofitted to existing. Fast Type 2 chargers are relatively small in size and can be placed in car parks in a similar manner as lighting columns. Dual socket chargers can be placed around the edge of car parks, each serving two spaces, with four socket chargers placed inbetween four spaces. If an EV can be charged at a slower rate, then the total power demand for the charge points installed in a car park can be reduced. Car parks will be mainly trickle charge top ups, or 8 hours for a full charge. Rapid full charges will be at transient locations, so allowing you to continue on your journey or to get home.

13.0 About the Author

So who am I and why am I writing about UK electric vehicle charging points? Well I am a Chartered Engineer by profession and work in the private sector, predominantly on public transportation systems. I have a keen interest in the environment, as well as technology. I have owned a Vauxhall Ampera Extended Range Electric Vehicle for over a year and was researching EVs for many months before making my purchase.

My choice of Ampera was dictated by the lack of electric vehicle charging points in the area where I live and in locations I would normally travel to. This meant my first choice of the Nissan Leaf was a no go, in terms of the travel I sometimes had to undertake. A Toyota Prius was considered, but I wanted a vehicle that could be recharged, which Toyota were not offering at the time, and I also wanted more than just a standard hybrid. My concern over the state of the current and expanding UK electric vehicle charging network and my quest to understand how I could fully utilise my Ampera as an electric vehicle, led me to the research and write this book.

I am also currently writing a book on my journey of researching electric vehicles, the difficulties I experienced with car dealerships and electric vehicle sales and my first year of ownership of an electric vehicle. This is due out in 2016 from Joe Public Publishing.

14.0　Acknowledgements

This book was not written in a vacuum and relied on input from various others over the last 18 months or so. This includes the following, for which I am most grateful to:

- Richard Bates for early encouragement with charge point infrastructure
- All those at the various charge point providers who responded to my various questions and queries in emails and telephone calls
- Everyone who allowed access to take pictures of charge points
- Paul Darlington, Senior Service Advisor, Porsche Centre, Bolton
- Paul Oxford from Autogas Ltd for allowing me a drive of their LPG converted Ford Focus and picking his brain on LPG conversions
- Mark Long for allowing me to detour his holiday so he could get me pictures of an ESB Ireland charge point
- Jordan Lewis, Product Genius, Halliwell Jones BMW, Warrington
- Derek Munro for proof reading and sense checking the text as well as providing encouragement and enthusiasm for all things EV

15.0　One Last Thing!

I hope you enjoyed reading this book as much as I enjoyed writing it. If you did enjoy the book please take a moment and leave a nice 5 star review on Amazon or any other website you may have purchased the book through. Then, please go and tell your friends, relatives and colleagues, all about this book!

If you have any constructive feedback on the book, or updated information, then please contact me via Twitter @ANHurst or via the Joe Public Publishing Facebook Page.

Thank you and happy electric motoring,

Andrew Hurst

Left: *To the east of the M1, between Junctions 18 and 19, and to the west of the village of Yelvertoft, lie two wind farms. Love them or hate them, these two wind farms of 13 turbines (8 + 5) are capable of producing some 26MW of green energy. On a bright April day in 2015, the Ampera took a detour to Manor Barns farm for a photo opportunity of a green car and green energy.*

Appendix A – Useful Websites

Online Mapping
- www.ChargeMap.com
- www.OpenChargeMap.org/app/
- www.PlugShare.com
- www.Zap–Map.com/location–search

Charge Point Schemes/Providers
- ChargePoint Genie   www.ChargePointGenie.com
- Chargernet   www.Poole.gov.uk/transport–and–streets/chargernet/
- Charge Your Car   www.ChargeYourCar.org.uk
- ecarNI – Northern Ireland   www.ecarNI.com
- Ecotricity   www.Ecotricity.co.uk
- Electric Corby   www.ElectricCorby.co.uk/projects/electric–vehicle–charging–infrastructure/
- Energise   www.EnergiseNetwork.co.uk
- ESB   www.esb.ie/electric-cars/index.jsp
- GMEV – Greater Manchester   www.ev.tfgm.com
- Greener Scotland   www.GreenerScotland.org
- Hackney   www.ChargePointGenie.com
- inmotion!   www.evinmotion.co.uk
- Just Park   www.JustPark.com
- Milton Keynes Crosslink   www.ChargeMasterPLC.com
- Plugged–in Midlands   www.PluggedInMidlands.co.uk
- POD Point   www.POD–Point.com
- Polar   www.PolarInstant.com
- Recharge   www.MerseyTravel.gov.uk/Recharge
- Source East   www.SourceEast.net
- Source London   www.SourceLondon.net
- Source West   www.SourceWest.info
- SSE   www.ChargePointGenie.com
- Tesla Supercharger Network www.TeslaMotors.com/en_GB
- Think Travel/Think Electric www.SourceWest.info
- Zero Carbon World   www.ZeroCarbonWorld.org

Charge Cables and Cable Bag Supplier
www.evConnectors.com

Electric Vehicle Compatibility Website
- www.UKevse.org.uk

Fuel Calculator

If you have not yet converted to an EV or wonder what you are spending on petrol in an your EREV, do the calculations on this website:

- www.fuel-economy.co.uk/calc.html

News Stories

The following news stories have been mentioned in this book:

New lithium-air battery design shows promise
- www.bbc.co.uk/news/science-environment-34669405

Ford reveals solar-powered car with sun-tracking technology
- www.bbc.co.uk/news/technology-25575306

First 'zero-emissions' hydrogen filling station opens
- www.bbc.co.uk/news/uk-england-south-yorkshire-34278051

How to produce hydrogen from algae
- www.bbc.co.uk/news/technology-34699166

Bio-Bus
- www.bbc.co.uk/news/uk-england-bristol-30115137
- www.geneco.uk.com/Biobug/biobus.aspx
- www.bbc.co.uk/news/uk-england-bristol-34919563
- www.firstgroup.com/bristol-bath-and-west/more/bio-b

E&OE

Appendix B – EV/EREV Vehicle Charge Socket List

Vehicle Type	EV Charge Socket 1	EV Charge Socket 2 (if fitted)
Audi A3 Sportback e-tron	Type 2	N/A
BMW i3 Electric Car	Type 2 (as standard) or Type 2/CCS Combined*	*Type 2/CCS Combined is an upgrade option
BMW i3 Range Extender	Type 2 (as standard) or Type 2/CCS Combined*	*Type 2/CCS Combined is an upgrade option
BMW i8 Range Extender	Type 2	N/A
Chevrolet Volt E-REV	Type 1	N/A
Citroen C-Zero Electric Car	Type 1	CHAdeMO
Ford Focus Electric Car	Type 1	N/A
KIA Soul EV Electric Car	Type 1	CHAdeMO
Mercedes-Benz C-Class Hybrid	Type 2	N/A
Mercedes-Benz S-Class Hybrid	Type 2	N/A
MIA Electric Electric	Type 1	N/A
Mitsubishi i-MiEV Electric Car Keiko	Type 1	CHAdeMO
Mitsubishi Outlander PHEV	Type 1	CHAdeMO
Nissan e-NV200 Electric Van	Type 1 (All models)	CHAdeMO (Only on Acenta Rapid and Tekna Models)
Nissan Leaf Electric Car	Type 1	CHAdeMO (Not all models – check before buying)

Vehicle Type	EV Charge Socket 1	EV Charge Socket 2 (if fitted)
Peugeot iOn Electric	Type 1	CHAdeMO
Peugeot Partner Electric Van	Type 1	CHAdeMO
Porsche 918 Spyder Hybrid	Type 2	N/A
Porsche Cayenne Hybrid	Type 2	N/A
Porsche Panamera Hybrid	Type 2	N/A
Renault Kangoo Z.E	Type 1	N/A
Renault Twizy Electric Car	Hard wired cable to 3 pin plug	N/A
Renault Zoe Electric Car	Type 2	N/A
Smart Fourtwo ED	Type 2	N/A
Tesla Model S	Type 2 Tesla	N/A
Tesla Roadster	Tesla Roadster Bespoke Socket	N/A
Toyota Prius Plug-In Hybrid	Type 1	N/A
Vauxhall Ampera	Type 1	N/A
Volkswagen e-Golf Electric Car	Type 2/CCS Combined	N/A
Volkswagen Golf GTE Plug-In Hybrid	Type 2	N/A
Volkswagen e-Up! Electric Car	Type 2/CCS Combined	N/A
Volvo V60 Plug-In Hybrid	Type 2	N/A
Volvo XC90 Hybrid	Type 2	N/A

Appendix C – Glossary of Terms

- AC – Alternating Current.
- Alternator – A device that when turned will generate an alternating current.
- Battery – A storage device that will hold DC electrical energy. Batteries manufactured from Lithium Ion (Li–ion) and are typically used to power an EV. Lead–acid batteries are used for secondary systems on an EV. Li–ion batteries are very expensive but they have a very high energy density, meaning they can store a lot of energy for their size.
- APT Technologies – Harrow based APT Technologies markets a number of charge point solutions under the "evolt" brand. These include both Type 2 fast and rapid tri-charger points. ChargePoint Genie/SSE, Greener Scotland/CYC and Recharge are three such organisations that use such charge points.
- CCS – Combined Charging System. This is a rapid DC charging backed by GM, Ford, Volkswagen, and BMW. It is also known as European Combined Charging System (CCS or 'Combo' 2). It entails a Type 2 style connector (but with only 3 pins – two signal pins and the earth) along with an additional 2 pin DC connector below this. This means the connector can only deliver DC charging though.
- CENEX – This is an acronym for Centre of Excellence for Low Carbon and Fuel Cell Technologies, based out of Loughborough University.
- Chargemaster PLC – Manufactures electric vehicle charge points and manages charge point schemes, through the brand name of Polar.
- Combo – See CCS.
- Combo 2 – See CCS.
- CHAdeMO – An abbreviation of "CHArge de MOve", equivalent to "charge for moving". The name is a pun for "cha demo ikaga desuka" which in Japanese, translates to English as "how about some tea?". This refers to the time it would take to charge a car (i.e. quite quickly!). The rapid DC charging standard is favoured by Nissan, Mitsubishi and Toyota. Also known as Japanese JARI/JEVS G105.
- CYC – Charge Your Car. An electric vehicle charge point scheme organisation originally formed as a joint venture between Charge Your Car (North) Ltd and Elecktromotive. Now a wholly owned subsidiary of Elecktromotive.
- DC – Direct Current.
- DBT-CEV – A French company manufacturing company of electric vehicle charge points. In the UK you will see their Quick Charger products in the guise of AC-DC rapid chargers used by Ecotricity (Type 2 and CHAdeMO)

and Universal, or tri-charger, used by Ecotricity and Chargemaster (Type 2, CCS and CHAdeMO). Other organisation may also use them.
- Elecktromotive – A UK based company founded in Brighton in 2003. They design, manufacture and sell charge–points under the name Elektrobay, as well as being the parent company for CYC.
- EREV or ER–EV – Extended Range Electric Vehicle. Usually an electric vehicle with a small petrol engine fitted to provide electricity to the motor when the battery is flat. The ICE drives an alternator to charge the battery and provide electricity to the electric motor. The ICE is not typically connected directly to the wheels and <u>the primary source of traction power to the wheels is the electric motor</u>. It is not the same as a Hybrid vehicle.
- ESB – Electricity Supply Board. An Irish electricity supply company.
- EV – Electric Vehicle. A vehicle with an electric motor and a battery to drive the wheels. To add confusion to the mix of EREVs and hybrids, some EV buses have a small ICE to provide power for heating only.
- EVSE – Electric Vehicle Supply Equipment. An official name for an electric vehicle charge point.
- Fast Charger – An AC charge point capable of charging most EVs in around 4 hours. Typically fitted with a type 2 socket and requires a charging cable.
- G105 – See CHAdeMO.
- GMEV – Greater Manchester Electric Vehicle. Charge point scheme for Greater Manchester area. Operated by CYC.
- Hybrid – This is a vehicle that is fitted with an <u>ICE as the primary source of traction</u> power driving the wheels. It is also fitted with a small electric motor that provides power at low speeds and when pulling away from a stop. It is not the same as an EREV.
- Hydrogen – The most abundant chemical element in the universe. The most recognisable source is water or H_2O, which covers around 70% of the planet. A water molecule is made up from 2x hydrogen atoms and 1x oxygen atom. By splitting a water molecule you can separate the individual atoms and harvest hydrogen gas and oxygen. However, the two hydrogen atoms connected to the oxygen atom are a very stable combination. To break these molecule bonds requires a lot of energy, hence why hydrogen production is very energy intensive. The hydrogen gas is compressed into a liquid and stored in a tank.
- ICE – Internal Combustion Engine. Usually petrol or sometimes diesel powered.
- IEC – International Electrotechnical Commission. This is the international standards and conformity assessment body for all fields of electro–technology

- IEC 62196–1 – This is a document that sets out the standards that are applicable to EV "charging accessories", namely: plugs, sockets and charge cables.
- Inverter – A device that will convert DC into AC.
- ITSO – Integrated Transport Smartcard Organisation Ltd. A not for profit organisation that is responsible for the technical standards and interoperability between various transport smartcard schemes that can be found on UK bus, tram and trains.
- Japanese JARI/JEVS G105 – See CHAdeMO.
- JARI – Japan Automobile Research Institute.
- JEVS – Japan Electric Vehicle Association Standard. Also see CHAdeMO.
- JEVS G105 – A JEVS technical standard that sets out what is basically the CHAdeMO connector. The standard's full name is JEVS G105–1993.
- LPG – Liquefied Petroleum Gas. A by-product of natural gas extraction and crude oil refining. It burns relatively cleanly producing no soot. It is compressed into a liquid and stored in a pressurised tank for use. It is sometimes refereed to a propane or butane, depending on the mix.
- Mennekes – A German manufacturer of industrial connectors, including what is known as the Type 2 EV connector.
- Mode 1 – Slow domestic AC Charging using a 3 pin plug. This is not used in the UK.
- Mode 2 – Slow domestic AC Charging via the built in control box (the plastic box fitted on the charge cable) and using a 3 pin plug. This is the norm for all EVs.
- Mode 3 – Fast or Rapid AC Charging from a charge point. This would include your professionally installed fast charge point at home or a charge point in the street or at the motorway services etc.
- Mode 4 – Rapid DC Charging from a charge point.
- OCM – Open Charge Map. A database developed by Zero Carbon World containing information on electric vehicle charging points across the world.
- OLEV – Office for Low Emission Vehicles. Part of the UK Government's Department for Transport, Department for Business, Innovation and Skills and Department of Energy and Climate Change
- PHEV - Plug-in Hybrid Electric Vehicle. An acronym used to distinguish the Mitsubishi Outlander EREV from the petrol/diesel variations of the same car.
- Polar – An electric vehicle charge point scheme organisation. A brand name of Chargemaster PLC.
- Rapid Charger – An AC or DC charge point capable of charging suitable EVs to 80% capacity in around 30 minutes. Typically fitted with a type 2

connector tethered cable for AC charging or a CCS or CHAdeMo connector on a tethered cable for DC charging.
- Rectifier – A device that will convert AC into DC.
- RFID – Radio Frequency Identification. Contactless smart card technology.
- REx – Range Extender. Another name for EREV. See EREV.
- SAE – Society of Automotive Engineers.
- SSE – Formerly known as Scottish and Southern Energy.
- TfGM – Transport for Greater Manchester.
- TfL – Transport for London.
- Traction Battery – See Battery.
- Tri–Charger – A single rapid charge point unit that has three tethered connectors providing a Type 2 AC connection, a rapid DC CHAdeMO connection and a rapid DC CCS connection.
- Triplex Rapid Charger – See Tri–Charger.
- Type 1 – SAE J1772 – This is a 5 pin EV charging connector, mainly used on EVs of American design. In North America it is sometimes referred to by its manufacturer's name of Yazaki. The 5 pins include earth, live and neutral as well as two control signal cables.
- Type 2 – This is a 7 pin EV charging connector often referred by the brand name of Mennekes. The 7 pins include earth, 3x live pins (for 3 phase AC charging) and neutral as well as two control signal pins.
- Type 2 Tesla – Similar to the Type 2 charging socket, but features some subtle variations in shape, including a keyway at the top. A standard Type 2 can still be used with the Model S.
- Type 4 – A CHAdeMO charger. See CHAdeMO.
- UltraCharger – A tri–charger designed and manufactured in the UK by Chargemaster.
- Siemens QC45 – This is a type of tri-charger (Type 2, CCS and CHAdeMO) that Siemens distribute in the UK. The charge point is actually a rebadged Efacec QC45 unit, which Siemens have the exclusive UK distribution rights for. Siemens operate a number of maintenance support contracts to councils for traffic light control systems, and so with the staff available on call for rapid response maintenance, adding charge points to their business portfolio made business sense for them. For Portuguese Efacec, it allows low risk sales opportunities into the UK market, with nationwide support services already in place. ESB Ireland and inmotion! are two organisations that use such charge points.
- Supercharger – A rapid DC charger by Tesla. Only suitable for the Tesla Model S car and capable of delivering 120kW of power.

Above: *Inconsistent signage is another issue of EV charge points. Is it recharging or charging? Point or station? Should there be words or just symbols? Or both?*

Printed in Great Britain
by Amazon